◎宁楠 著

Java 零基础实战

人民邮电出版社

北京

图书在版编目（CIP）数据

Java零基础实战 / 宁楠著. -- 北京：人民邮电出版社，2019.5
ISBN 978-7-115-50895-9

Ⅰ. ①J… Ⅱ. ①宁… Ⅲ. ①JAVA语言—程序设计 Ⅳ. ①TP312.8

中国版本图书馆CIP数据核字(2019)第037865号

内 容 提 要

Java 是一门经典的面向对象的编程语言，具有很强的通用性、平台可移植性和安全性，并且一直在编程语言排行榜上稳居前列。本书旨在通过浅显易懂的语言，讲清 Java 的特性，帮助读者掌握面向对象编程的核心思想；同时，通过丰富实用的代码示例，帮助读者快速上手，轻松掌握 Java 语言。

本书分 11 章，带领读者从零开始认识 Java，了解 Java，会用并且用好 Java。书中不仅对 Java 的运行机制、体系结构和基本的安装配置进行了讲解，还对面向对象编程的思想进行了深刻的剖析和总结，同时作者以丰富的代码示例演示了基本的 Java 编程。除此之外，本书还对 Java 的集合框架、多线程、实用类、IO 流、反射机制和 Web 编程等进行了讲解，让读者在掌握基本编程技巧的基础上，进一步探索 Java 的强大功能。

本书由资深 Java 开发老手编写，汇集了丰富的实践经验和实用的编程思想，非常适合想入门 Java 的新手学习，也适合用作计算机相关专业的辅助教程，其他非科班出身的想学习 Java 编程的读者也可以参考学习。

◆ 著　　宁　楠
责任编辑　胡俊英
责任印制　焦志炜

◆ 人民邮电出版社出版发行　北京市丰台区成寿寺路 11 号
邮编 100164　电子邮件 315@ptpress.com.cn
网址 http://www.ptpress.com.cn
固安县铭成印刷有限公司印刷

◆ 开本：787×1092　1/16
印张：21.5　　　　　　　2019 年 5 月第 1 版
字数：505 千字　　　　　2024 年 7 月河北第 3 次印刷

定价：69.80 元

读者服务热线：(010)81055410　印装质量热线：(010)81055316
反盗版热线：(010)81055315
广告经营许可证：京东市监广登字20170147号

序言

Java 编程学习之道

我大学读的是软件工程专业，在校期间学校开设了很多高级编程语言课程，如 C、C++、C#、Java 等。经过一段时间的学习实践和对比，我决定专攻 Java，原因有三点：一、Java 易上手且功能强大，经过 20 多年的发展，Java 生态体系已经非常成熟；二、应用面广，无论是 Web 端、移动端，还是企业级项目，都可以使用 Java 进行开发；三、Java 开发工程师的市场需求量很大，并且待遇也不错。基于以上三点原因，我非常坚定地选择了 Java 作为自己主要的编程语言。当然这并不是说其他语言不好，每种语言都有自己的优势和劣势，没有绝对的好与坏，要结合自己的实际情况选择一门编程语言学习。在我看来，对于零基础的初学者，选择相对容易上手且成熟的编程语言来入门，效率会更高，做到"先有再精"，后期可根据工作情况和个人选择来学习其他的高级编程语言，有了基础以后，学起来就会很快。编程思想都是相通的，拿我本人来说，除了继续深入学习 Java，我也在学习其他编程技术，毕竟选择从事编程行业的工作就注定了要不断地学习。

我从第一次接触 Java 到现在也有将近 10 年的时间了，从初级开发工程师做起，一路进阶到高级开发、项目经理、技术总监。在职业生涯中的每个阶段，我对 Java 编程都会有新的认识，甚至很多时候对于过去的理解有颠覆性的改变。这是因为学习技术需要有一个沉淀和积累的过程，由量变完成质变。毕竟，写过一万行代码和写过 10 万行代码的人对技术的认知是不一样的，不同阶段对编程的理解也是不一样的，我们在工作和学习中需要不断总结思考，通过实际项目研发去夯实基础，建立自己的知识体系。

对于初入职场的新人来说，什么是最重要的？不是理论知识是否扎实，更不是你懂多少新技术，而是解决问题的能力，公司花钱雇你来上班，那你最基本的职责就是要在规定的时间内完成规定的任务。项目研发的不可预期性很高，你不知道什么时候会遇到哪些技术难点，解决这些难点需要花费多少时间和精力，这就要求你具备快速解决问题的能力。能在短时间内把问题搞定，能保证项目按时交付，你才是一名合格的程序员。在工作中，我们常常会遇到痴迷于技术的程序员，这本身没有问题，每个程序员都应该痴迷于技术，但是需要明确的是技术应为项目服务，也就是说能按时交付合格的项目是最重要的。应用的技术是否高端并不是最重要的，如果不能按时完成项目，连最基本的要求都无法达到，何谈技术优化升级？所以我们在实际工作中不要过于追求技术，一切要以实践为主。我曾经遇到过这样的同事，技术很厉害，喜欢在项目中炫技，把自认为好的技术强行加入项目中。他不考虑是否合适，是否会影响项目进度，想方设法也要把自认为好的技术加进去，最终的结果往往是辛苦加班很多天也没有实现需要的功能，而别人使用相对传统的技术很快就能完成任务。这就是本末倒置，得不偿失。技术固然很重要，但是应该以项目为重，为项目选择最合适的技术，而不是你认为最新最好的技术。公司对项目的要求首先是能按时交付，在这个基础之上，再去考虑技术的迭代和优化。

看到这里，有的人可能会说，那编程工作就很简单了，只需要掌握一些基本的技术，能完成工作就可以了，不需要学习什么新技术呀。我上面说的是对一个开发者最基本的要求，或者说是最低标准，如果你希望自己的事业节节高升，就需要不断学习新技术，不断更新优化自己的知识体系，并且记住一点，那就是只学有用的。什么是有用的呢？一方面是你当前工作所需要具备的技能，可以更好地帮你完成工作。另一方面是前瞻性的技术，比如未来两三年你可能需要掌握的，或者是大趋势所指向的技术。

当你从初级开发者成长为中高级开发者时，就不能仅限于实现基本的业务功能了，这样的工作没有多少含金量，干一年和干 3 年不会有太大的区别。打个比方，当搬砖对你来说已经是驾轻就熟，你就需要去思考怎么设计房子了，也就是从 CRUD 基本操作到软件设计架构的进阶。怎样提高自己的软件架构能力呢？首先你要具备扎实的基础知识，其次要有足够的项目经验，还要视野开阔，在技术领域的涉猎面要广。提高编程能力最直接也有效的方法就是看源码，学习源码是有一定门槛的，刚开始看的时候可能会遇到很多问题，或者根本就看不懂。这个时候也不能放弃，要逼着自己看源码，第一遍第二遍完全不懂没关系，坚持看到第三遍、第四遍的时候会发现自己好像懂了那么一点，继续坚持下去，你就会理解源码的思路了。

学习的阶段性很重要，找准自己当前所处的位置，学自己最应该学习的东西，慢慢提升自己的水平，不要跟风，不要别人学什么你就学什么。不同的阶段需要有针对性地部署学习计划来不断地提高和完善自己。假如你是刚入行的初级开发者，你首先应该考虑如何提高自己解决问题的能力，学习业务知识，更加高效地完成工作，而不是去学习软件架构设计。连地基都没有，何来万丈高楼？学习编程是一个从无到有、从有到精、从精到广的过程，想要做好 Java 软件开发的工作，需要不断地总结、思考、学习。

欢迎大家加入 QQ 群 "Java 零基础实战"（群号：688588534）来参与互动学习，我会在群里答疑解惑，还会适时分享实用且超值的学习资料。

宁楠

2019 年 1 月于西安

致 谢

感谢一起工作过的同事们，感谢你们在工作中给予我的帮助。感谢人民邮电出版社胡俊英编辑，她非常专业的指导和建议使得本书的内容更加完善。感谢我的家人，因为有你们的理解和支持，我才能顺利完成本书的写作。

作者简介

宁楠，资深 Java 开发工程师，历任高级开发工程师、项目经理、技术总监。同时也是公众号原创博主、知识星球嘉宾、慕课网讲师、GitChat 认证作者，对 Java 编程有着丰富的经验和独到的见解，热衷于分享技术干货。欢迎大家搜索微信公众号"Java 大联盟"关注我的技术文章，也期待与大家有更多的技术交流。

前言

写作本书的目的

我个人在学习 Java 的过程中有这样一种感受，很多书本上的知识偏重于理论，没有太多的实践案例，不利于快速上手。网上的一些博客教程偏重于实践，没有太多晦涩难懂的理论知识，能快速上手进行开发，往往学习效率更高。但是博客也有不好的地方，一方面是不够系统，往往只是一些单一的技术点，并没有形成完整的知识体系。另一方面是不够严谨，错误也比较多，找到一篇优质且技能点完整的博客不是一件容易的事情。那么一本以实践为主，理论为辅，能快速上手的 Java 系统性入门图书，对于零基础的初学者来讲就显得尤为重要，甚至可以帮助他们达到事半功倍的效果。我写作本书的目的就是为所有对 Java 感兴趣的零基础读者提供一本可以帮助他们快速入门，以实际开发为导向的书。

一直以来我都有做笔记和写总结的习惯，并且坚持以实践为主，用实践去验证理论。也正是因此，我的笔记和总结实用性很强，我通过各种网络渠道分享过一些自认为比较好的笔记，反响还不错，尤其是很多初学者很喜欢看我写的教程，觉得通俗易懂，很适合他们。我在工作之余录制过一些视频课程，也做过在线直播授课，久而久之发现初学者的很多问题和困惑都是相同的，就集中在那么几个关键点上，所以我在写教程的时候就会侧重于这些技术点的讲解，并且尽可能用通俗易懂的语言去讲解，抓住了关键点，自然会得到受众群体的青睐。

我将多年来的心得体会和技术笔记进行汇总，历时数月，历经反复推敲修改，最终整理成本书。目标群体定位为零基础的初学者，让他们可以通过本书实现 Java 编程的快速入门，并且能将书中所讲的知识学以致用，写出功能完善的 Java 项目。

读者对象

本书适合以下几类读者阅读：

- 计算机相关专业的学生；
- 有兴趣从事 Java 开发工作的人群；
- 初级 Java 开发工程师。

如何阅读本书

本书的结构是按照 Java 知识体系的进阶来编写的，全书分为 4 个部分共 11 章。

第 1 部分是 Java 基础，也就是第 1~3 章，介绍了 Java 的历史、功能、运行机制和体系结构，以及 Java 环境的安装。重点讲解了 Java 的基本语法、变量、内存模型、运算符、流程控制、循环和数组等内容。

第 2 部分是 Java 面向对象，包括第 4~5 章，主要介绍了面向对象编程思想，以及在 Java 中的应用，包括类与对象、封装、继承、抽象、多态、包装类和异常。

第 3 部分是 Java 高级应用，包括第 6~9 章，主要讲解了 Java 的集合框架、多线程、实用类和 IO 流，学完这部分内容，就可以使用 Java 开发出功能完善的程序了。

第 4 部分是底层扩展，包括第 10~11 章，主要讲解了 Java 的反射机制和网络编程，这两块内容对于初学者可能有些难度，但是都非常重要。要搞清楚 Java 企业级开发组件和框架的底层原理，这两部分知识必不可少。

大家可以根据自己的需求选择阅读的侧重点，不过初学者最好能够按照顺序来阅读，这样可以有一个循序渐进的过程，对整个技术结构有一个清晰的梳理，有助于读者建立自己的知识体系。

本书示例代码说明

本书共有 300 多段示例代码，为了让阅读更加精简，书中只摘取了代码的核心部分，其余部分做了省略，省略的部分包括包的引入、导入其他包的类、成员变量的 getter、setter 方法以及 try-catch 等常规操作，完整的代码片段如下所示。

```java
package com.southwind.io;
import java.io.FileOutputStream;
import java.io.IOException;
import java.io.OutputStream;
public class IOTest {
    private String path;
    public String getPath() {
        return path;
    }
    public void setPath(String path) {
        this.path = path;
    }
    public static void main(String[] args) {
        OutputStream outputStream = null;
        try {
            outputStream = new FileOutputStream("...");
            outputStream.write(99);
```

```
      } catch (IOException e) {
        // TODO Auto-generated catch block
        e.printStackTrace();
      } finally {
        try {
          outputStream.close();
        } catch (IOException e) {
          e.printStackTrace();
        }
      }
    }
  }
```

精简之后的代码片段如下所示。

```
public class IOTest {
  private String path;
  //getter、setter 方法

  public static void main(String[] args) {
    OutputStream outputStream = null;
    try {
      outputStream = new FileOutputStream("...");
      outputStream.write(99);
    } catch (IOexception e) {
    } finally {
      try {
        outputStream.close();
      } catch (IOException e) {
      }
    }
  }
}
```

读者在复现书中示例时需要注意补全代码，完整的源码可到异步社区的官网下载。

关于勘误

作者水平有限，加之时间仓促。虽然我花费了大量的精力反复审读，但书中仍难免会有一些错误和纰漏。如果读者发现了任何问题，恳请反馈给我，可以通过邮箱 ningnan9801@163.com 与我取得联系，也可以提交到异步社区。

资源与支持

本书由异步社区出品，社区（https://www.epubit.com/）为您提供相关资源和后续服务。

配套资源

本书提供配套代码资源，要获得该配套资源，请在异步社区本书页面中单击 ，跳转到下载界面，按提示进行操作即可。注意：为保证购书读者的权益，该操作会给出相关提示，要求输入提取码进行验证。

如果您是教师，希望获得教学配套资源，请在社区本书页面中直接联系本书的责任编辑。

提交勘误

作者和编辑尽最大努力来确保书中内容的准确性，但难免会存在疏漏。欢迎您将发现的问题反馈给我们，帮助我们提升图书的质量。

当您发现错误时，请登录异步社区，按书名搜索，进入本书页面，单击"提交勘误"，输入勘误信息，单击"提交"按钮即可。本书的作者和编辑会对您提交的勘误进行审核，确认并接受后，您将获赠异步社区的100积分。积分可用于在异步社区兑换优惠券、样书或奖品。

扫码关注本书

扫描下方二维码，您将会在异步社区微信服务号中看到本书信息及相关的服务提示。

与我们联系

我们的联系邮箱是 contact@epubit.com.cn。

如果您对本书有任何疑问或建议，请您发邮件给我们，并请在邮件标题中注明本书书名，以便我们更高效地做出反馈。

如果您有兴趣出版图书、录制教学视频，或者参与图书翻译、技术审校等工作，可以发邮件给我们；有意出版图书的作者也可以到异步社区在线提交投稿（直接访问 www.epubit.com/selfpublish/submission 即可）。

如果您是学校、培训机构或企业，想批量购买本书或异步社区出版的其他图书，也可以发邮件给我们。

如果您在网上发现有针对异步社区出品图书的各种形式的盗版行为，包括对图书全部或部分内容的非授权传播，请您将怀疑有侵权行为的链接发邮件给我们。您的这一举动是对作者权益的保护，也是我们持续为您提供有价值的内容的动力之源。

关于异步社区和异步图书

"异步社区"是人民邮电出版社旗下IT专业图书社区，致力于出版精品IT技术图书和相关学习产品，为作译者提供优质出版服务。异步社区创办于2015年8月，提供大量精品IT技术图书和电子书，以及高品质技术文章和视频课程。更多详情请访问异步社区官网 https://www.epubit.com。

"异步图书"是由异步社区编辑团队策划出版的精品IT专业图书的品牌，依托于人民邮电出版社近30年的计算机图书出版积累和专业编辑团队，相关图书在封面上印有异步图书的 LOGO。异步图书的出版领域包括软件开发、大数据、AI、测试、前端、网络技术等。

异步社区

微信服务号

目 录

第 1 部分　Java 基础

第 1 章　Java 初体验 2
- 1.1　Java 概述 2
  - 1.1.1　什么是 Java 2
  - 1.1.2　Java 的运行机制 4
  - 1.1.3　Java 三大体系 5
- 1.2　搭建 Java 开发环境 6
  - 1.2.1　安装配置 Java10 6
  - 1.2.2　Java 程序开发步骤 11
- 1.3　小结 13

第 2 章　Java 入门 14
- 2.1　开发第一个 Java 程序 14
  - 2.1.1　使用 Eclipse 开发程序 14
  - 2.1.2　编码规范 21
  - 2.1.3　注释 24
  - 2.1.4　关键字 27
- 2.2　变量 29
  - 2.2.1　什么是变量 29
  - 2.2.2　如何使用变量 30
- 2.3　基本数据类型 32
- 2.4　数据类型转换 33
  - 2.4.1　自动转换 33
  - 2.4.2　强制转换 35
- 2.5　运算符 36
  - 2.5.1　赋值运算符 36
  - 2.5.2　基本算术运算符 37
  - 2.5.3　复合算术运算符 40

- 2.5.4　关系运算符 41
- 2.5.5　逻辑运算符 43
- 2.5.6　条件运算符 46
- 2.5.7　位运算符 47
- 2.6　小结 50

第 3 章　Java 进阶 51
- 3.1　流程控制 51
  - 3.1.1　if-else 51
  - 3.1.2　多重 if 53
  - 3.1.3　if 嵌套 55
  - 3.1.4　switch-case 56
- 3.2　循环 58
  - 3.2.1　while 循环 58
  - 3.2.2　do-while 循环 61
  - 3.2.3　for 循环 63
  - 3.2.4　while、do-while 和
    for 这 3 种循环的区别 64
  - 3.2.5　双重循环 65
  - 3.2.6　终止循环 69
- 3.3　数组 70
  - 3.3.1　什么是数组 70
  - 3.3.2　数组的基本要素 71
  - 3.3.3　如何使用数组 71
  - 3.3.4　数组的常用操作及
    方法 74
  - 3.3.5　二维数组 78
- 3.4　综合练习 80
- 3.5　小结 86

第 2 部分　Java 面向对象

第 4 章　面向对象基础 88
- 4.1　什么是面向对象 88
- 4.2　类与对象 89
  - 4.2.1　类与对象的关系 89
  - 4.2.2　定义类 90
  - 4.2.3　构造函数 91
  - 4.2.4　创建对象 92
  - 4.2.5　使用对象 93
  - 4.2.6　this 关键字 93
  - 4.2.7　方法重载 94
  - 4.2.8　成员变量和局部变量 95
- 4.3　封装 98
  - 4.3.1　什么是封装 98
  - 4.3.2　封装的步骤 99
  - 4.3.3　static 关键字 101
- 4.4　继承 105
  - 4.4.1　什么是继承 105
  - 4.4.2　子类访问父类 106
  - 4.4.3　子类访问权限 109
  - 4.4.4　方法重写 111
  - 4.4.5　方法重写 VS 方法重载 114
- 4.5　多态 114
  - 4.5.1　什么是多态 114
  - 4.5.2　多态的使用 117
  - 4.5.3　抽象方法和抽象类 119
- 4.6　小结 122

第 5 章　面向对象进阶 123
- 5.1　Object 类 123
  - 5.1.1　认识 Object 类 123
  - 5.1.2　重写 Object 类的方法 125
- 5.2　包装类 131
  - 5.2.1　什么是包装类 131
  - 5.2.2　装箱与拆箱 132
- 5.3　接口 135
  - 5.3.1　什么是接口 135
  - 5.3.2　如何使用接口 136
  - 5.3.3　面向接口编程的实际应用 138
- 5.4　异常 142
  - 5.4.1　什么是异常 142
  - 5.4.2　异常的使用 142
  - 5.4.3　异常类 146
  - 5.4.4　throw 和 throws 147
  - 5.4.5　自定义异常类 151
- 5.5　综合练习 152
- 5.6　小结 157

第 3 部分　Java 高级应用

第 6 章　多线程 160
- 6.1　进程与线程 160
- 6.2　Java 中线程的使用 163
  - 6.2.1　继承 Thread 类 163
  - 6.2.2　实现 Runnable 接口 165
  - 6.2.3　线程的状态 166
- 6.3　线程调度 167
  - 6.3.1　线程休眠 167
  - 6.3.2　线程合并 169
  - 6.3.3　线程礼让 171
  - 6.3.4　线程中断 173
- 6.4　线程同步 175
  - 6.4.1　线程同步的实现 175
  - 6.4.2　线程安全的单例模式 183
  - 6.4.3　死锁 187
  - 6.4.4　重入锁 189
  - 6.4.5　生产者消费者模式 194
- 6.5　综合练习 196
- 6.6　小结 198

第 7 章　集合框架 199
- 7.1　集合的概念 199
- 7.2　Collection 接口 200
  - 7.2.1　Collection 接口的定义 200

7.2.2 Collection 的子接口 201
7.3 List 接口 201
 7.3.1 List 接口的定义 201
 7.3.2 List 接口的实现类 202
7.4 Set 接口 207
 7.4.1 Set 接口的定义 207
 7.4.2 Set 接口的实现类 208
7.5 Map 接口 214
 7.5.1 Map 接口的定义 214
 7.5.2 Map 接口的实现类 215
7.6 Collections 工具类 220
7.7 泛型 222
 7.7.1 泛型的概念 222
 7.7.2 泛型的应用 224
 7.7.3 泛型通配符 226
 7.7.4 泛型上限和下限 227
 7.7.5 泛型接口 228
7.8 综合练习 229
7.9 小结 232

第 8 章 实用类 233

8.1 枚举 233
8.2 Math 236
8.3 Random 237
8.4 String 238
 8.4.1 String 实例化 238
 8.4.2 String 常用方法 242
8.5 StringBuffer 244
8.6 日期类 246
 8.6.1 Date 246
 8.6.2 Calendar 248
8.7 小结 249

第 9 章 IO 流 250

9.1 File 类 250
9.2 字节流 251
9.3 字符流 257
9.4 处理流 265

9.5 缓冲流 267
 9.5.1 输入缓冲流 268
 9.5.2 输出缓冲流 274
9.6 序列化和反序列化 279
 9.6.1 序列化 279
 9.6.2 反序列化 280
9.7 小结 281

第 4 部分 底层扩展

第 10 章 反射 284

10.1 Class 类 284
10.2 获取类结构 287
 10.2.1 获取类的接口 288
 10.2.2 获取父类 289
 10.2.3 获取构造函数 290
 10.2.4 获取方法 292
 10.2.5 获取成员变量 294
10.3 反射的应用 296
 10.3.1 反射调用方法 296
 10.3.2 反射访问成员变量 299
 10.3.3 反射调用构造函数 302
10.4 动态代理 303
10.5 小结 309

第 11 章 网络编程 310

11.1 IP 与端口 312
 11.1.1 IP 312
 11.1.2 端口 314
11.2 URL 和 URLConnection 314
 11.2.1 URL 314
 11.2.2 URLConnection 317
11.3 TCP 协议 318
11.4 UDP 协议 322
11.5 多线程下的网络编程 324
11.6 综合练习 326
11.7 小结 328

第1部分 Java 基础

第1章 Java 初体验

> Hello，欢迎来到 Java 的世界！本书适用于零基础且有兴趣学习 Java 编程的小伙伴，全程会以理论和实践相结合的方式，手把手地教你用 Java 开发出自己的程序。你是否已经迫不及待了呢？现在就随我一起来探索 Java 的学习之道吧！
>
> 本章我会带领大家一起来认识 Java 这门高级编程语言，在学习一个新技术时，应该从 3 个方面入手。首先，它是什么？其次，它能干什么？最后，如何使用？大家需要记住这 3 个步骤，我们后续的讲解都是按照这个模式进行的。好了，接下来就开始我们的 Java 初体验。

1.1 Java 概述

本节带领大家了解 Java 的基本知识、运行原理、体系结构、Java 的用途，以及如何使用 Java 进行编程。

1.1.1 什么是 Java

众所周知，IT 是一个高速发展、技术更迭日新月异的行业。随着编程技术的不断发展、更新，越来越多的编程语言层出不穷，令大家不知如何选择。面对种类繁多的编程语言，零基础的初学者究竟应该选择哪种编程语言来入门呢？掌握哪门编程语言能够更好地应对 IT 技术日益革新、多态化发展的大趋势呢？综合以上思路，这门语言应该具备以下特质：

- 容易上手；
- 适用于多平台，多行业；
- 发展稳定；
- 开发需求量大。

什么语言这么强大，可以满足这些要求呢？没错，就是 Java。Java 是由 Sun（Stanford

University Network）公司于 1995 年 5 月 23 日正式推出的一套计算机高级编程语言，它拥有 20 多年的发展历史，这足以表明 Java 很稳定。

Java 适用于多个领域，从早期的终端设备到现在流行的电商、桌面管理软件、机顶盒设备、车载导航、安卓移动端等，涵盖面非常之广，并且可以做到跨平台，无论是 Windows 系统、Linux 系统，还是 Mac OS 系统，都可以运行 Java 程序。

多年以来，Java 始终在编程语言排行榜中名列前茅（见图 1-1），是全球范围内使用人数最多的编程语言之一。同时，市场对 Java 开发工程师的需求量也很大，在各大招聘网站上，Java 相关岗位的招聘信息非常多。

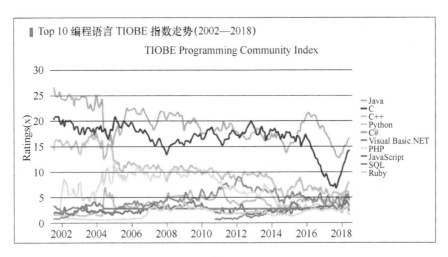

图 1-1

Java 是面向对象的编程语言，面向对象是一种易理解、上手快的编程思想，同时 Java 提供了功能非常强大的系统类库，多种多样的工具类任你使用，开发者只需要关注业务逻辑，然后根据具体需求去调用这些工具类即可，底层的具体实现不需要关注。因此，相比于 C++这种偏向底层的高级编程语言，Java 更容易上手。

Java 语言具备如下特点。

1．简洁高效

Java 语言非常简洁，相比于 C++中头文件、指针等各种抽象的概念，Java 更好理解、便于上手，同时还提供了功能强大的系统类库，使开发变得更加简洁高效。

2．跨平台

程序在不同平台的兼容性问题一直困扰着开发者，如果我们开发的程序能够无障碍地同时运行在 Windows、Mac OS 和 Linux 系统中，那是多么美妙的一件事。Java 就帮我们实现了这个美好的愿望，一套代码可以在多种平台上运行。

3．面向对象

面向对象是一种编程思想，这种编程思想的诞生，对于软件工程有着划时代的意义。开发者告别了面向过程开发的烦琐步骤，从一个新的维度重新解读编程这件事，极大地提升了软件开发效率和能力，Java 就是这样一种面向对象的高级编程语言。

4．分布式计算

Java 提供了一套网络操作类库，很适合开发分布式计算的程序，开发者可以通过调用类库进行网络程序开发，实现分布式特性。

5．健壮性

Java 提供了非常强大的排错机制，在程序编译阶段就可以检测出程序中的错误，无需等到运行时才暴露出存在的问题。同时在运行阶段会再一次进行相应的检查，多种手段保证了程序的稳定性和健壮性。

6．可处理多线程

线程是进程的基本单位，是程序开发中必不可少的一种基础资料，Java 提供了良好的多线程处理机制，使程序具备更为优秀的交互性。

好了，听完了这些，你是不是已经迫不及待地想要开始学习 Java 了呢？别着急，我们说过，学习新技术，首先应该了解它是什么？其次要了解它能干什么，接下来，我们就来一起看看 Java 有哪些方面的实际应用。

在全球范围内，选择 Java 作为后台开发语言的公司数不胜数，比较知名的国外互联网公司包括 Google、YouTube、Amazon、Twitter 等，国内的知名公司有阿里、腾讯、百度、新浪、搜狐等。尤其是在电商领域，大部分公司都在使用 Java，包括天猫、京东、苏宁易购、当当、美团等，这么多知名企业都选择 Java 作为开发语言，你还在犹豫什么呢？

1.1.2　Java 的运行机制

简单来讲，Java 开发可分为 3 步：

- 在后缀为.java 的文件中编写 Java 程序，此文件称之为 Java 源文件；
- 通过编译器将源文件编译为后缀为.class 的字节码文件；
- 计算机读取字节码文件运行程序。

运行原理如图 1-2 所示。

图 1-2

我们可以这样理解，Java 源文件是开发者编写的，以开发者自己能看懂的方式去编写代码，但是计算机无法直接识别编写好的程序，因为计算机只能识别二进制的数据，相当于两者语言不通，要进行交流就必须有翻译，编译器就是这个翻译，它可以将开发者编写的程序翻译成计算机能识别的二进制数据，即将 Java 源文件编译为字节码文件，这样一来计算机就可以运行程序了。

　　Java 程序并不是运行在计算机底层的，Java 拥有自己的虚拟计算机，这个虚拟的计算机有自己的内存，有自己的磁盘，我们把它叫作 Java 虚拟机（Java Virtual Machine，JVM）。所有的 Java 程序都是运行在 JVM 上的，正是因为有了 JVM 这样一种机制，Java 程序才能做到跨平台，不同的操作系统只要可以安装 JVM，就可以运行 Java 程序。JVM 可以将不同操作系统的底层运行机制进行屏蔽，读取与平台无关的字节码文件，由 Java 解释器将 JVM 的程序运行在不同的平台上。编译好的字节码文件只需要识别 JVM，而不需要关心更底层的操作系统，由 JVM 去适应并识别不同的操作系统，如图 1-3 所示。

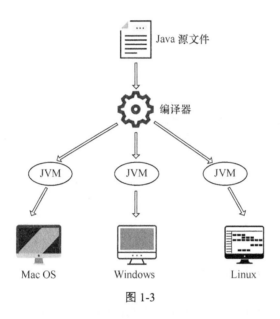

图 1-3

　　举一个生活中的例子，一个中国商人，需要同时跟来自美国、意大利和德国的 3 个客户洽谈合作，但是中国商人只会说汉语，所以他高价雇佣了一个同时精通英语、意大利语、德语的高级翻译，帮他把汉语分别翻译给来自美国、意大利和德国的客户，这样就可以无障碍地沟通并完成合作了。

　　在这个例子中，中国商人相当于 Java 源程序，高级翻译相当于 JVM，而来自美国、意大利、德国的客户相当于不同的操作系统。

1.1.3　Java 三大体系

　　Java 语言后来衍生出 3 个体系分支，分别是 J2SE、J2ME、J2EE。

　　J2SE（Java2 Platform Standard Edition）定义了 Java 的核心类库，包含了各种常用组件，是 Java 开发的基础。

　　J2ME（Java2 Platform Micro Edition）是基于 J2SE 衍生出的专用于移动设备的开发组件，如手机、机顶盒、车载导航等。

J2EE（Java2 Platform Enterprise Edition）是基于 J2SE 扩展出的企业级开发组件，提供了 Java Web 相关的开发组件，如 Servlet、JSP 等，是 Java 开发的主流技术。

2005 年之后，三大体系被重新命名，其中 J2SE 更名为 Java SE，J2ME 更名为 Java ME，J2EE 更名为 Java EE。在三大体系中，Java SE 是核心，Java ME 和 Java EE 是在 Java SE 的基础上发展起来的，如图 1-4 所示。

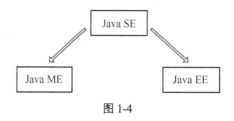

图 1-4

1.2 搭建 Java 开发环境

上一节我们学习了 Java 的基本概念，了解了 Java 的运行原理。Java 程序必须依赖于 JVM 才能运行，所以开发 Java 程序的第一步就是在计算机上安装 Java 环境，本节就带大家一起来安装和配置 Java 环境。

1.2.1 安装配置 Java10

Java 经历了这么多年的发展，最新的版本是 Java10（作者在编写本书时的最新版），我们就以 Java10 为例，给大家演示如何安装和配置 Java 环境。首先我们需要了解什么是 Java 环境，有两个重要的概念我们需要掌握——JRE 和 JDK。

JRE（Java Runtime Environment）是 Java 的运行环境，包括 JVM 和 Java 基础类库，一台计算机要运行 Java 程序，就必须有 JRE。

JDK（Java Development Kit）是 Java 开发包，它包含 JRE 和编译 Java 源文件的编译器，我们要在一台计算机上进行 Java 程序开发，就必须安装 JDK。

由此可知，JRE 是 Java 程序运行环境，JDK 是 Java 程序开发环境，而 JDK 包含了 JRE，我们只需要安装 JDK 即可。

1. 下载 JDK

（1）打开 Java 官方网站，选择"JDK Download"下载 JDK10.0.2（作者在编写本书时的最新版），如图 1-5 所示。

（2）在打开的新网页选中"Accept License Agreement"选项，如图 1-6 所示。

图 1-5

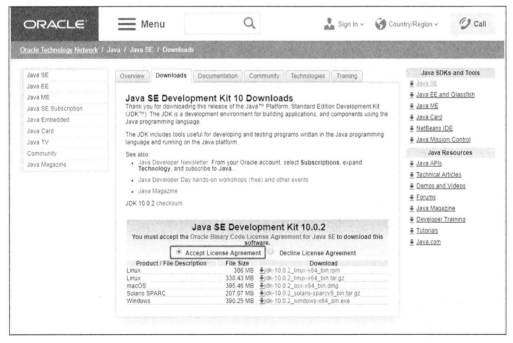

图 1-6

（3）根据你的操作系统选择对应的 JDK 安装文件，这里我们选择 Windows 系统下的 JDK，如图 1-7 所示。

图 1-7

（4）下载好的安装文件如图 1-8 所示。

图 1-8

2．安装 JDK

（1）找到 JDK 安装文件，双击鼠标左键运行该文件。

（2）一直单击"下一步"按钮，全部选择"默认"即可。

3．配置环境变量

（1）配置 path 环境变量使计算机在运行程序时可以找到 Java 程序的路径。

（2）配置 classth 可以设置 class 文件的路径信息。

配置环境变量的具体步骤如下所示。

（1）找到自己安装 Java 的路径，复制下来留着下一步备用，如图 1-9 所示。

（2）计算机→属性→高级系统设置→高级→环境变量→系统变量，点击"新建"，为变量起名"JAVA_HOME"，变量值设置为上一步复制的路径，如图 1-10 所示。

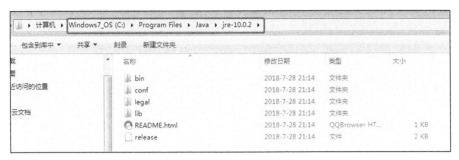

图 1-9

图 1-10

(3) 在 Path 路径的开头添加 "%JAVA_HOME%\bin;",如图 1-11 所示。

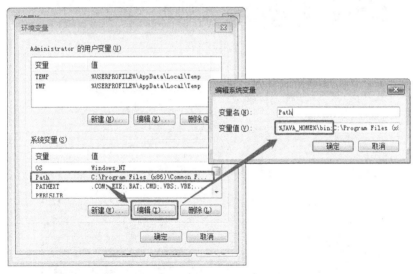

图 1-11

4. 检测是否安装成功

(1) 按 Windows +R 组合键,输入 cmd,单击"确定"按钮打开终端,如图 1-12 所示。

图 1-12

（2）在终端输入"javac"以及"java"，如果安装成功就可以看到如图 1-13 和图 1-14 所示的信息。

图 1-13

图 1-14

（3）若出现"javac/java 不是内部命令"字样，则表示环境配置失败，需要重新配置。

（4）输入"java -version"，可以查看 Java 的版本信息，如图 1-15 所示。

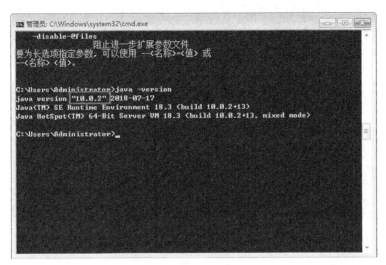

图 1-15

1.2.2 Java 程序开发步骤

想必读到这里，大家都已经成功地配置好了 Java 环境，准备工作就绪，接下来我们就可以正式开始写代码了。

我们的第一个 Java 程序是输出"Hello World"。在 D 盘新建一个文件夹，命名为 java，进入 java 文件夹，新建一个名为 HelloWorld.java 的文件，这就是 Java 源文件。接下来，用记事本或者其他编辑器打开文件来编写代码，如代码 1-1 所示。

代码 1-1

```java
public class HelloWorld{
    public static void main(String[] args) {
        System.out.println("Hello World");
    }
}
```

我们之前介绍过，Java 程序开发分三步：第一步编写，第二步编译，第三步运行。代码 1-1 编写完成之后，我们需要进行第二步：编译。

（1）打开 CMD 终端（按 Windows+R 组合键，然后输入"cmd"），用"d:"命令进入计算机的 D 盘，然后通过 cd 命令进入 Java 程序所在的文件夹，如图 1-16 所示，HelloWorld.java 保存在 D 盘的 java 文件夹中。

（2）如图 1-17 所示，使用 javac HelloWorld.java 命令编译 HelloWorld.java 文件。

（3）如图 1-18 所示，在编译成功后，会看到 D:/java 文件夹中自动生成了一个 HelloWorld.class 文件，该文件就是我们之前提到的字节码文件，JVM 就是通过读取这个文件来运行程序的。

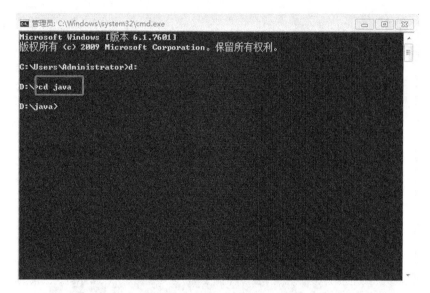

图 1-16

图 1-17

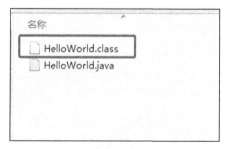

图 1-18

（4）如图 1-19 所示，使用 java HelloWorld 命令来运行 Java 程序，计算机打印输出了"HelloWorld"字符串。

图 1-19

成功！这里对代码进行简单的说明，public static void main(String[] args)是程序的入口，我们希望计算机执行的代码都写在这里，大家可以先这样理解，后面我们会详细讲解 main 方法。

注意事项：
- 编译程序的命令是 javac；
- 运行程序的命令是 java；
- 编写代码时注意字母的大小写，Java 程序对大小写敏感。

1.3 小结

作为本书的开篇内容，本章首先为大家介绍了 Java 的基本运行原理、体系结构以及 Java 能开发哪些应用程序，让初学者对 Java 有一个基本的认识和概念。介绍完 Java 的基本概念，接下来为大家讲解如何安装配置 Java 环境，第一步安装 JDK，第二步配置环境变量，完成之后就可以在计算机上开发 Java 程序了。Java 程序开发共分为 3 个步骤：第一步编写 Java 源代码，第二步将 Java 源代码编译成 JVM 能识别的字节码文件，第三步计算机读取字节码文件并运行程序。

我们在学习一个新知识的时候，首先是要了解它是什么，它能做什么，然后才是怎么使用。本章对这 3 个问题做了详细解答，为后续的深入学习做好了铺垫。

第 2 章　Java 入门

通过第 1 章的学习，想必大家已经对 Java 语言有了初步的认识，对 Java 的发展历程、运行原理和环境安装有了一定的了解。在本章中，我们继续来学习 Java 的详细语法与开发规范，教大家用 Java 编写出简单的程序。

2.1 开发第一个 Java 程序

2.1.1 使用 Eclipse 开发程序

还记得上一章内容讲到过的 Java 开发三步骤吗？第一步，编写 Java 源代码；第二步，将 Java 源代码编译成字节码文件；第三步，运行程序。这三步是 Java 开发编译的底层步骤，在真正的项目开发中，如果我们按照这种方式去开发代码，很显然是非常麻烦的一件事情。为了简化开发步骤，提高开发效率，我们需要使用集成开发环境（Integrated Development Environment，IDE）来编写代码。

Java 开发常用的 IDE 有 Eclipse、NetBeans、IntelliJ IDEA。使用 NetBeans 的人较少，Eclipse 是当下主流的集成开发环境，IntelliJ IDEA 也是一款优秀的 IDE，具有代码智能提示等强大的功能。因为目前选择 Eclipse 的公司居多，所以本书采用 Eclipse 作为开发环境。

使用 Eclipse 集成环境进行开发，可以将原来的三步走简化为两步：首先编写代码，然后直接运行即可。但底层实际还是三步，只是 Eclipse 对编译和运行进行了整合，开发者只需要进行一次操作即可完成编译并同时看到运行结果。另外，Eclipse 还有代码提示功能，可提高我们的编码效率，你是否已经跃跃欲试了呢？

图 2-1

1．安装 Eclipse

（1）进入 Eclipse 官网，下载安装文件，如图 2-1 所示。

（2）运行安装文件，如图 2-2 所示。

图 2-2

(3)在弹出的窗口选择第二项,该版本包含了 JavaEE 开发组件,如图 2-3 所示。

图 2-3

(4)选择安装目录,单击"install"进行安装,如图 2-4 所示。

(5)安装完成,双击"eclipse.exe",打开 Eclipse,在弹出的菜单选择工作空间,即代码在本地硬盘的保存路径,如图 2-5 所示。

(6)看到如图 2-6 所示的界面,表示我们已经安装成功,接下来就可以使用 Eclipse 编写程序了。

图 2-4

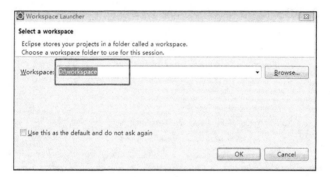

图 2-5

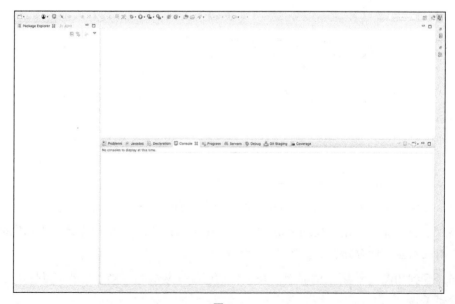

图 2-6

2. 工程

一个项目是由很多 Java 文件组成的，我们用工程（Project）来管理所有的 Java 文件。

（1）在 Eclipse 中新建 "Java Project"，单击 File→New→Java Project，如图 2-7 所示。

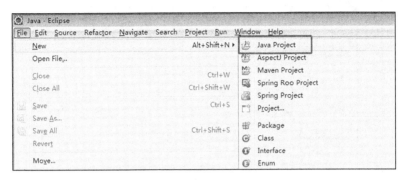

图 2-7

（2）在弹出窗口的 "Project name" 处填写自定义工程名，Eclipse 会自动关联本机的 Java 环境，也可以手动修改，单击 "Finish" 完成创建，如图 2-8 所示。

图 2-8

（3）创建成功，工程结构如图 2-9 所示，src 目录下保存了开发者自定义的 Java 文件。JRE System Library 是 Eclipse 自动导入的 Java 程序系统库，我们在编写程序时可以调用系统类库来完成相关功能。

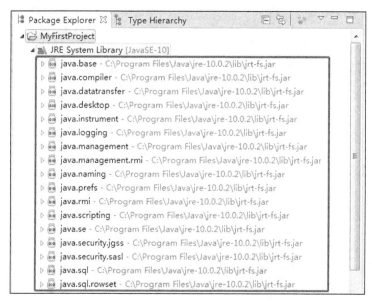

图 2-9

（4）工程创建好了，开始写代码，我们之前说过 Java 是通过类（Class）来组织代码结构的，所以首先创建一个 Class，在 src 目录处单击鼠标右键→New→Class，如图 2-10 所示。

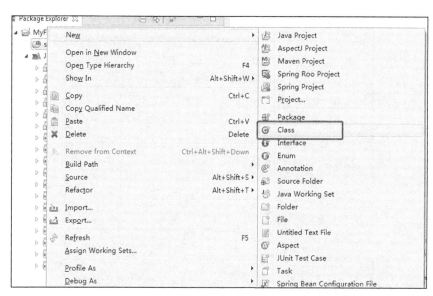

图 2-10

（5）在弹出窗口的"Name"处输入 Class 名称，也就是 Java 文件名，单击"Finish"完成，如图 2-11 所示。

（6）创建完成之后，src 目录下会生成 HelloWorld.java 文件，存放在默认包（default package）中。至于什么是包，我们会在后面的章节中详细讲解。Eclipse 会自动打开

HelloWorld.java 文件，同时自动生成 class 相关代码，我们只需在 class 内部添加逻辑代码即可，如图 2-12 所示。

图 2-11

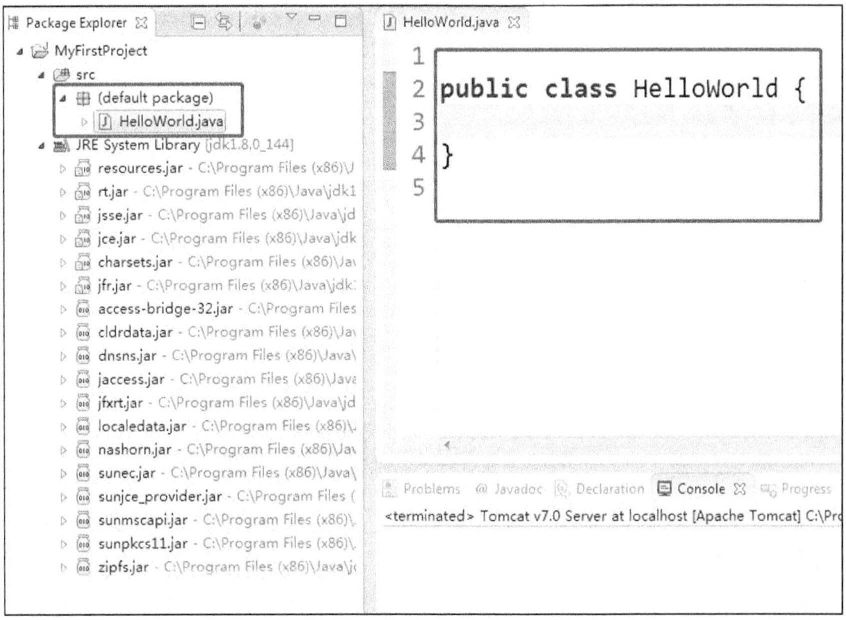

图 2-12

（7）在 HelloWorld.java 中编写打印"HelloWorld"的代码，如代码 2-1 所示。

代码 2-1
```
public class HelloWorld{
   public static void main(String[] args) {
      System.out.println("Hello World");
   }
}
```

（8）在 HelloWorld.java 空白处单击鼠标右键选择→Run As→Java Application，如图 2-13 所示。

图 2-13

（9）在控制台（Console）可以看到打印结果，一个简单的 Java 程序就编写并运行完毕了，如图 2-14 所示。

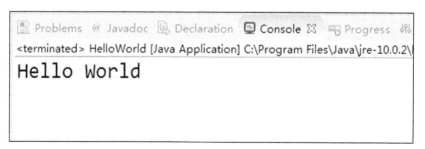

图 2-14

（10）虽然我们可以直接运行程序并看到结果，但实际上 Eclipse 还是经过了先编译，再运行的过程，只不过我们看不到编译的过程，Eclipse 将编译好的字节码文件保存在工程的 bin 目录中，在 Eclipse 中我们看不到 bin 目录，但它是真实存在的，JVM 运行的就是 bin 目录下的字节码文件，如图 2-15 所示。

图 2-15

2.1.2 编码规范

我们在编写 Java 程序时，需要注意编码规范。无规矩不成方圆，Java 要求我们必须按照一定的规范去写代码。一部分是强制性的，必须执行，不按规范写程序无法运行；另一部分是非强制性的建议规范，即不按照规范写并不影响程序的运行，只是为了统一标准，提高代码的可读性，在团队协作开发时，程序员们能更好地读懂彼此的代码。编码规范的内容很多，我们先介绍现阶段需要掌握的编码规范，在后续的章节中会继续介绍相应的编码规范。

1. 强制编码规范

（1）Java 程序的文件名与类名必须一致，若不一致，Java 源文件将无法编译通过。

例：把 HelloWorld.java 文件中的类名改为 MyHelloWorld，如代码 2-2 所示。

代码 2-2
```java
public class MyHelloWorld{
   public static void main(String[] args) {
      System.out.println("Hello World");
   }
}
```

现在我们使用 dos 命令对 HelloWorld.java 进行编译，此时类名和文件名不一致，会报错，如图 2-16 所示。

```
D:\workspace\MyFirstProject\src>javac HelloWorld.java
HelloWorld.java:2: 错误: 类MyHelloWorld是公共的, 应在名为 MyHelloWorld.java 的文件中声明
public class MyHelloWorld {
       ^
1 个错误

D:\workspace\MyFirstProject\src>
```

图 2-16

Eclipse 有自动提示错误信息的功能，如果在 Eclipse 中修改类名，Eclipse 会直接给出相应提示，HelloWorld.java 文件左下角有一个标志，同时具体代码处会有一个标志。当我们把鼠标移动至类名处时，会自动弹出错误信息及解决方案，第一种解决方法，将文件名改为 MyHelloWorld.java；第二种解决方案，将类名改为 HelloWorld。无论哪种方案，

都是让类名与文件名保持一致，如图 2-17 所示。

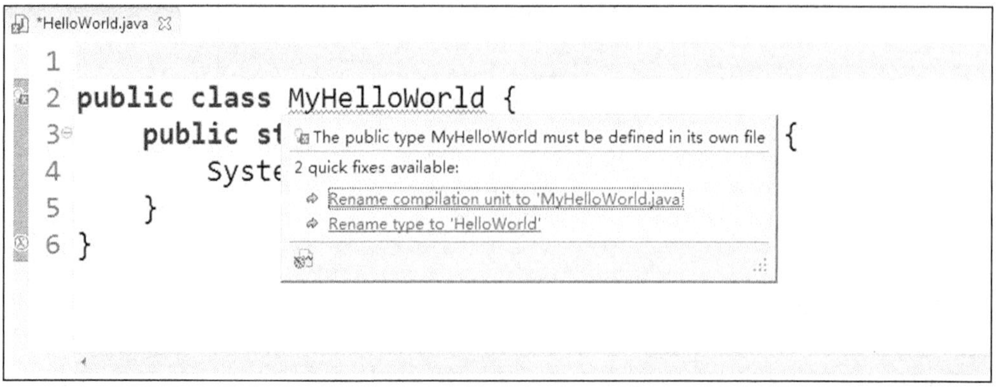

图 2-17

我们可以直接单击两种解决方法中的任意一个，Eclipse 会自动帮我们完成修改文件名或修改类名的工作。怎么样，是不是很智能？其实 Eclipse 还有很多非常实用的功能，学会使用 IDE 工具，可以提高我们的开发效率，在后面的章节中也会给大家陆续介绍 Eclipse 的更多实用功能。

（2）main 方法是程序的入口，方法的定义必须严格按照格式书写，public static void main(String[] args){}，缺一不可，否则 Java 程序无法运行，只有参数列表的形参名称可自定义，如代码 2-3 所示。

代码 2-3

```
public class HelloWorld{
    public static void main(String[] parameter) {
        System.out.println("Hello World");
    }
}
```

（3）类是组织 Java 代码结构的，类中的方法是具体执行业务逻辑的，无论是类还是方法，都必须使用花括号{}来组织其结构，并且必须成对出现。如果缺少某个花括号，Eclipse 会自动提示错误信息，HelloWorld.java 文件左下角有一个标志，同时具体代码错误处会有标志，如图 2-18 所示。

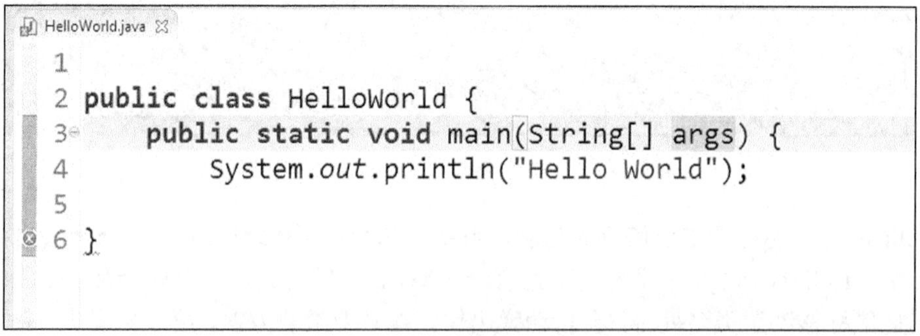

图 2-18

当我们把鼠标移动至 ⊗ 处，Eclipse 会自动弹出具体错误信息，要求补全花括号，如图 2-19 所示。

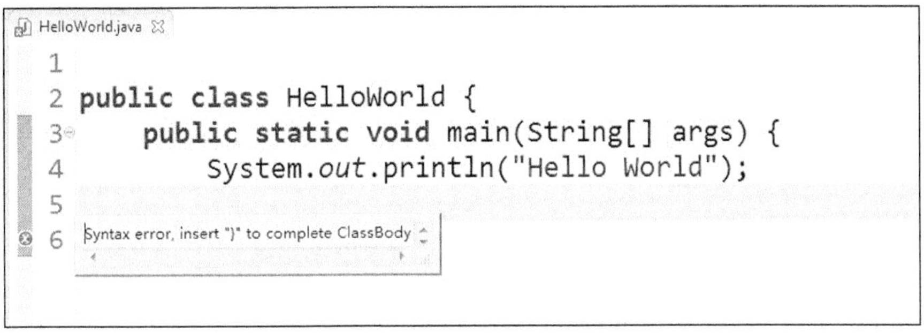

```
HelloWorld.java
1
2  public class HelloWorld {
3      public static void main(String[] args) {
4          System.out.println("Hello World");
5
6  Syntax error, insert "}" to complete ClassBody
```

图 2-19

2．建议编码规范

（1）通常情况下，一行只写一条语句，程序结构清晰，方便阅读者理解代码，如代码 2-4 所示。

代码 2-4

```
public class HelloWorld {
   public static void main(String[] args) {
      System.out.println("Hello World");
      System.out.println("I love Java");
      System.out.println("你好");
   }
}
```

反例如代码 2-5 所示。

代码 2-5

```
public class HelloWorld {public static void main(String[] args) {
     System.out.println("Hello World");System.out.println("I love Java");System.out.println("你好");}
}
```

可以看出，结构混乱，可读性也很差。

（2）遵守一行只写一条语句的同时，还需要注意代码缩进，不要每一行代码都顶着左边写，这个规范同样是为了提高代码的可读性，反例如代码 2-6 所示。

代码 2-6

```
public class HelloWorld {
public static void main(String[] args) {
System.out.println("Hello World");
System.out.println("I love Java");
System.out.println("你好");
}
}
```

这样的代码没有层次结构，没有体现出类和方法的分层结构关系。类是代码的外层结构，顶着左边写，方法是类内部执行具体业务逻辑的模块，应该距离左边 4 个空格，按下键盘的 tab 键，可自动完成 4 个空格的缩进。同理，方法中的代码属于方法的下层结构，应该距离方法的左边界 4 个空格，如代码 2-7 所示。

代码 2-7

```
public class HelloWorld {
    public static void main(String[] args) {
        System.out.println("Hello World");
        System.out.println("I love Java");
        System.out.println("你好");
    }
}
```

2.1.3 注释

实际开发中项目通常是由一个团队来协同完成的，每个程序员的编程思路是不一样的，编码风格也大相径庭，代码量较大时就会产生一些问题，比如程序员 A 写完了自己的代码，程序员 B 需要调用，但是他发现自己读懂程序员 A 的代码就需要花费大量的时间，这样就会大大降低开发效率，如何解决这个问题呢？我们可以通过给代码添加注释的方式来帮助开发者更好地读懂代码。

Java 注释就是用通俗易懂的语言对代码进行描述解释，方便自己和他人阅读，以达到快速、准确地理解代码的目的。注释可以是编程思路，也可以是功能描述或者程序的作用，总之就是对代码的进一步阐述。

Java 代码中的注释是不会被编译的，也就是说计算机不会读取注释，而是直接跳过，注释是专门写给开发者看的。

1. Java 中的注释

（1）单行注释：//注释内容

（2）多行注释：/*注释内容

　　　　　　　　注释内容

　　　　　　　　注释内容 */

（3）文档注释：/**注释内容

　　　　　　　　*注释内容

　　　　　　　　*注释内容

　　　　　　　　*/

文档注释可以使用 JDK 的 javadoc 工具来生成信息，并输出到 HTML 文件中，可以更加方便地记录程序信息，javadoc 注释标签语法如表 2-1 所示。

表 2-1

标签	描述
@author	标识作者
@deprecated	标识过期的类或成员
@exception	标识抛出的异常
@param	标识方法的参数
@return	标识方法的返回值
@see	标识指定参考的内容
@serial	标识序列化属性
@serialData	标识通过 writeObject( ) 和 writeExternal( )方法写的数据
@serialField	标识 ObjectStreamField 组件
@since	标识引入一个特定的变化
@throws	标识抛出的异常
@version	标识版本
{@docRoot}	标识当前文档根目录的路径
{@inheritDoc}	标识从直接父类继承的注释
{@link}	标识插入的主题链接
{@linkplain}	标识插入的主题链接,但是该链接显示纯文本字体
{@value}	标识常量的值,该常量必须是 static 属性

2. Eclipse 中注释用的快捷键

(1)单行注释:添加注释,选中目标代码,按 **Ctrl+/**组合键即可,如图 2-20 所示。如果要删除注释,选中目标代码,按 **Ctrl+/**组合键即可。

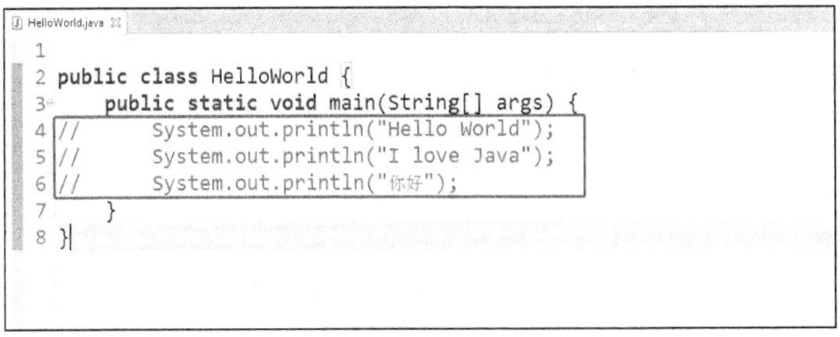

图 2-20

(2)多行注释:添加注释,选中目标代码,按 **Ctrl+Shift+/**组合键即可,如图 2-21 所示。如果要删除此类注释,没有快捷键,需要手动删除。

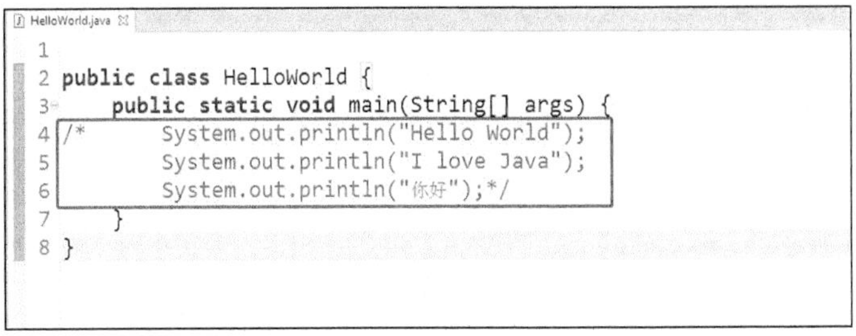

图 2-21

（3）文档注释：添加注释，按/**+Enter 组合键会自动生成注释，如图 2-22 所示。而删除此类注释则没有快捷键，需要手动删除。

图 2-22

单行注释和多行注释一般用于方法体中具体的业务代码，而文档注释一般用于描述类或者方法，写在类定义处或者方法定义处。描述类时，会自动生成@author 标签来描述作者，作者名即计算机的用户名，如图 2-23 所示。

图 2-23

描述方法，会自动根据方法的参数列表，返回值类型生成@param 标签和@return 标签，若方法的返回值为 viod，即没有返回值，也就不会生成@return 标签，如图 2-24 所示。

图 2-24

在自动生成标签的基础上，我们可以手动添加更多的注释来更加详细地描述目标类或目标方法，并且在添加标签时，Eclipse 会有代码提示，可通过按 Alt+？组合键来选择，如图 2-25 所示。

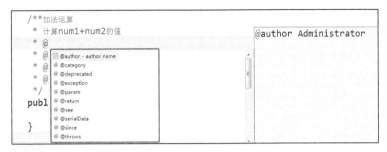

图 2-25

添加标签完成之后，如图 2-26 所示。

图 2-26

2.1.4 关键字

1. 关键字的定义

Java 关键字是指 Java 语言预先定义好的，有指定意义的标识符，是程序的核心组成。不是由开发者来定义的，简单来说 Java 程序就是"关键字+Java 类库+开发者自定义的标识符"。关键字来构建程序的核心骨架，Java 类库提供程序要调用的方法，开发者自定义的标识符来指定程序完成相关工作。例如在代码 2-8 中，pubilc、class、static、void 为关键字；main、System.out.println 为 Java 类库方法（main 是一个特殊方法，表示主线程）；HelloWorld、"Hello World"、"I love Java"、"你好"就是开发者自定义的标识符了。关键字搭建了程序的骨架，我们希望程序输出的词组通过类库方法来完成打印工作。

代码 2-8

```java
public class HelloWorld {
    public static void main(String[] args) {
        System.out.println("Hello World");
        System.out.println("I love Java");
        System.out.println("你好");
    }
}
```

Java 关键字可以表示一个基本数据类型，可以表示流程控制，可以作为类和方法的修饰符等，但是不能作为包名、类名、方法名、参数名、变量名，这些概念在后续的章节都会介绍。Java 关键字全部为小写，在 Eclipse 环境中，字体会自动加粗加黑，如图 2-27 所示，方框标记的都是关键字。

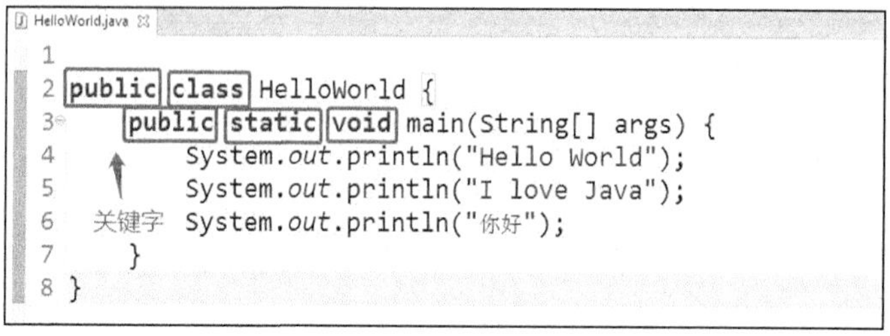

图 2-27

2．关键字的含义

Java 关键字的含义如表 2-2 所示。

表 2-2

关键字	含义
abstract	标识类或方法为抽象类型
assert	断言、用于调试程序
boolean	基本数据类型
break	跳出整个循环体
byte	基本数据类型
case	与 switch 搭配使用，标识变量符合某个条件
catch	与 try 搭配使用，捕获某个异常
char	基本数据类型
class	标识一个类
const	保留字
continue	跳出当前循环，执行下一次循环
default	与 switch 搭配使用，表示默认值
do	与 while 搭配使用，标识循环体的开始
double	基本数据类型
else	与 if 搭配使用，表示条件不成立
enum	枚举类型
extends	标识类的继承
final	修饰常量，表示类不能被继承，方法不能被覆盖，变量值不能被修改
finally	与 try、catch 搭配使用，表示无论如何都会执行
float	基本数据类型
for	标识循环
goto	保留字
if	条件判断
implements	标识类实现某个接口
import	在类中引入某个包中的类

续表

关键字	含义
instanceof	判断变量的数据类型
int	基本数据类型
interface	接口
long	基本数据类型
native	修饰本地方法，表示该方法不是由 Java 实现的
new	创建实例对象
package	表示包，是一种组织 Java 类的结构
private	访问权限修饰符，表示私有
protected	访问权限修饰符，表示保护
public	访问权限修饰符，表示公有
return	在方法中返回一个值
short	基本数据类型
static	静态修饰符
strictfp	表示 FP-strict，精确浮点
super	表示调用父类的成员
switch	流程控制语句
synchronized	线程同步
this	表示调用当前实例对象的成员
throw	抛出异常
throws	标识在方法中可能会抛出的异常
transient	表示不可序列化状态
try	标识可能会抛出异常的代码块
void	表示当前方法没有返回值
volatile	修饰变量，线程在访问该变量时，读取的是该变量被修改之后的值，保证每个线程读到最新值
while	标识循环

2.2 变量

2.2.1 什么是变量

变量是计算机语言中的一个概念，可以表示某个具体数值，并且这个值可以改变，所以叫变量。与之对应的是常量，常量也是用来表示某个数值的，但值是固定的，不能改变。为什么要使用变量呢？我们通过一个生活中的例子来解释这个问题：你用 100 元本金购买了一款理财产品，年利率是 3%，那么年底你的收益是多少？

这是一道非常简单的数学题，收益=100×3%=3 元。这个过程是在我们的大脑中进行运算的，大脑会首先记住 100 和 3%这两个值，然后做乘法运算。现在让程序去完成这个运算，其实和我们大脑计算的步骤是一样的：在计算机中存储 100 和 3%这两个数据，然后进行乘法运算。那么首先要搞清楚的是，计算机是如何存储数据的呢？

计算机存储数据的地方叫作内存，内存会为不同的数据开辟不同的空间来存储，所以数据和数据之间是相互独立的，互不影响。类似于超市的储物柜，我们把不同的数据保存在不同的柜子中，如图 2-28 所示。

图 2-28

好了，现在已经把数据存储在内存中了，接下来进行运算。运算时需要把这两个数据从内存中取出来，怎么取呢？每一个内存空间都有自己独一无二的地址，程序就是通过内存地址找到具体的内存空间，从中取出数据。相当于我们去超市买东西，把自己的包存在储物柜中，每个储物柜都有自己的编号，离开时通过编号找到储物柜，再把包取出来，计算机在内存中存取数据就是类似这样的过程。

内存地址是十六进制的数据，例如 0x6fff5cde3d6c。这样的表达式不便记忆，如果一个运算中需要用到多个数据，我们就需要记住每一个内存的十六进制地址，然后分别取出来。这种方法非常麻烦，不利于我们编写程序，那有没有解决方案呢？有的话，怎么解决？就是使用变量。

变量是一个概念，存储在内存中，方便存取内存中的数据，是程序中存储数据的基本单元。变量有三要素：数据类型、变量名、变量值，数据类型是变量中保存的值的数据类型，不同的数据需要用不同的类型来保存，整数、小数、文本信息、日期等都需要用不同的数据类型来表示。不同的数据类型需要开辟的内存空间也是不同的，相当于储物柜有大小之分，分别存放大小不同的物品。

变量名是由开发者自己定义的，上面说到内存地址不好记，那么既然计算机分配给你的数据你记不住，那你就自己取名字吧，自己取的名字总能记得住吧。所以我们就可以给 100 和 3% 分别取名 principal 和 interest，内存地址不好记忆的问题迎刃而解。变量值就是内存中存储的数据，principal 的值就是 100，interest 的值就是 3%。综上所述，变量是一种概念，内存是具体实现变量这一概念的物理空间。

2.2.2 如何使用变量

接下来我们学习变量的创建过程。

（1）声明变量的数据类型和变量名，计算机根据数据类型在内存中开辟相应大小的空间，同时变量名的定义要符合规则，可以包含数字、字母、下划线、$，不能包含空格、运算符，不能用纯关键字命名，不能以数字开头，大小写字母可混用，首字母应小写，后续单词的首字母应大写，如 userId、studentName 等。

（2）给内存空间赋值，该值就是变量值，如代码 2-9 所示。

代码 2-9

```java
public class HelloWorld {
    public static void main(String[] args) {
        //1.开辟内存空间，100 是整数，数据类型为 int
        int principal;
        //2.赋值
        principal = 100;
    }
}
```

当 "int principal;" 这行代码运行完成之后，计算机会在内存中开辟一块存放 int 类型数据的空间，如图 2-29 所示。

图 2-29

当 "principal = 100;" 这行代码运行完成之后，程序会将 100 放入内存，如图 2-30 所示。

图 2-30

以上就是变量的创建过程，我们也可以将两行代码整合成一行，如代码 2-10 所示。

代码 2-10

```java
public class HelloWorld {
    public static void main(String[] args) {
        //开辟内存空间+赋值
        int principal = 100;
    }
}
```

2.3 基本数据类型

Java 共有 8 种基本数据类型，程序中常用的数据类型有整数、小数、字母、单词、汉字等。我们可以将这些数据分为两大类，数值类型（整数、小数）和非数值类型（字母、单词、汉字），基本数据类型的具体信息如表 2-3 所示。

表 2-3

分类	基本数据类型	所占空间	描述
数值类型	byte	1 字节（8 位）	数据的最小单位，一字节为 8 位二进制数，所以 byte 的取值范围是–128~127
	int	4 字节（32 位）	整数的最常用类型，取值范围是–2147483648~2147483647
	short	2 字节（16 位）	短整型，特定情况下使用 short 比使用 int 可以节省很多内存，取值范围–32768~32767
	long	8 字节（64 位）	长整型，描述 int 无法承载的海量数据，取值范围 –9223372036854775808~9223372036854775807
	float	4 字节（32 位）	单精度浮点型，float 类型的数据末尾必须添加 "f" 或 "F"，用以区分 double 类型，取值范围 3.402823e+38 ~ 1.401298e-45
	double	8 字节（64 位）	双精度浮点型，比 float 存储范围更大，精度更高，取值范围 1.797693e+308~ 4.9000000e-324
非数值类型	char	2 字节（16 位）	表示单个字符，可以是字母、汉字、数字、符号等，存储范围\u0000~\uFFFF
	boolean	1/8 字节（1 位）	判定逻辑条件，只有两个值 true 和 false，true 表示成立，false 表示不成立

在实际开发中，我们常用的基本数据类型有 int、double、boolean。在描述文本类型的数据时，char 有很大的局限性，因为只能表示单个字符。如果希望输出一段文本，很显然使用 char 就不合适了，这种情况下，我们通常要使用 String 来表述文本数据。String 是 JDK 提供给开发者调用的一个类，其本质是一个 char 类型数组，即由多个 char 类型数据组成的一个数据，所以也叫字符串。在后面的章节我们会详细讲解 String 类，这里大家做一个简单的了解即可。

接下来我们看一个例子，用程序输出用户信息：编号 1，姓名张三，性别男，身高 176cm，体重 60.5kg，要求使用变量定义用户的各项基本信息并打印输出，结果如图 2-31 所示。

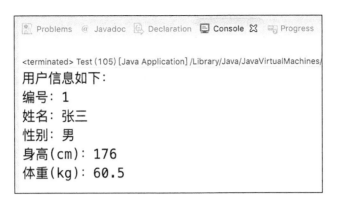

图 2-31

这个需求很简单，用户一共有 5 项基本信息：编号、姓名、性别、身高、体重。我们需要创建 5 个变量来分别表示这 5 项数据。首先要选择数据类型，姓名为"张三"，8 种基本数据类型中只有 char 可以用来表示文本，但是 char 只能表示一个汉字，所以这里我们需要用到 String 来定义姓名，需要注意的是 String 不是基本数据类型。编号 1 和身高 176 是整数，很显然使用 int，性别"男"可以使用 char，体重是 60.5，不是整数，所以我们选择 double。核心问题解决了，接下来就是将这 5 个变量进行打印输出，如代码 2-11 所示。

代码 2-11

```
public class Test {
    public static void main(String[] args) {
        int id = 1;
        String name = "张三";
        char gender = '男';
        int height = 176;
        double weight = 60.5;
        System.out.println("用户信息如下：");
        System.out.println("编号："+id);
        //打印其他属性值
    }
}
```

需要注意的是"System.out.println("编号："+id);"中的"+"作用是将左右两边的数值以文本的形式进行拼接，即将"编号："和 id 变量的值进行拼接，所以结果就是"编号：1"。

2.4　数据类型转换

2.4.1　自动转换

我们知道每个变量都有自己特定的数据类型，那么如果一个数值跟变量的数据类型不匹配，我们能不能将该数值赋值给变量呢？答案是可以的。因为 Java 中的变量可以进行数据类型转换，如我们可以把一个整数赋给 double 类型的变量，如代码 2-12 所示。

代码 2-12

```
public class Test {
    public static void main(String[] args) {
        int num = 10;
        double num2 = num;
        System.out.println(num2);
    }
}
```

运行结果如图 2-32 所示。

```
@ Javadoc  Declaration  Console ⊠  Progress  Debug
<terminated> Test (141) [Java Application] C:\Program Files (x86)\Java\
10.0
```

图 2-32

通过结果我们可以看到，打印的结果是 10.0，即整数 10 现在已经变成了小数 10.0。虽然从结果来看数值并没有改变，但在计算机底层是两种完全不同的数据类型。由原来的 int 类型变为 double 类型，这一过程叫作自动类型转换，程序会自动去识别数据类型并完成转换。这种操作对参与的数据类型是有要求的，并不是所有的数据类型之间都可以完成自动转换，只能由低字节向高字节进行转换，反之则不行。double 类型的存储空间为 8 字节，int 类型的存储空间为 4 字节，所以 int 类型可以自动转为 double 类型，double 类型不能自动转为 int 类型。

同时，浮点型（double、float）描述数值的精度比整型（short、int、long）高，或者说浮点型可以对数值进行更精确的描述，所以整型数据类型也可以自动转为浮点型，即使 long 为 8 字节，float 为 4 字节，不满足只能由低字节向高字节进行转换的原则，也同样可以完成自动转换。我们说的自动类型转换只包括所有数值类型，不包括 char 和 boolean。基本数据类型自动转换的关系为：byte→short→int→long→float→double，如代码 2-13 所示。

代码 2-13

```
public class Test {
    public static void main(String[] args) {
        byte num1 = 10;
        short num2 = num1;
        int num3 = num2;
        long num4 = num3;
        float num5 = num4;
        double num6 = num5;
        System.out.println(num1);
        //依次打印 num2、num3、num4、num5、num6
    }
}
```

运行结果如图 2-33 所示。

```
@ Javadoc  Declaration  Console ⊠  Progress  Debug
<terminated> Test (141) [Java Application] C:\Program Files (x86)\Java\
10
10
10
10
10.0
10.0
```

图 2-33

2.4.2 强制转换

上一节提到 int 类型可以自动转为 double 类型，而 double 类型不能自动转为 int 类型。这句话其实是不严谨的，严格意义上来讲是错的。即 double 类型是可以转为 int 类型的，只不过程序不会自动完成，需要我们手动干预进行强制转换。如何操作呢？很简单，如代码 2-14 所示。

代码 2-14

```
public class Test {
    public static void main(String[] args) {
        double num = 10.5;
        int num2 = (int) num;
        System.out.println(num2);
    }
}
```

运行结果如图 2-34 所示。

图 2-34

通过结果可以看到 double 在转为 int 时会造成精度损失，由 10.5 变成了 10，小数点后全部丢弃，将浮点型数值强转为整型时，只保留小数点左边的数值，不会按照四舍五入的方式来完成整数位的进位。使用 Eclipse 进行编程，遇到不能自动转换数据类型的情况时，Eclipse 会自动提示错误并给出解决方案，如图 2-35 所示。

1. Add cast to 'int'：进行强制类型转换，点击之后会自动补全强制类型转换的代码。
2. Change type of 'num2' to 'double'：将 num2 的数据类型改为 double。
3. Change type of 'num' to 'int'：将 num 的数据类型改为 int。

第 2 项和第 3 项都是将变量的数据类型进行修改,使得赋值变量和被赋值变量的数据类型保持一致,也可以解决不能自动转换的问题。

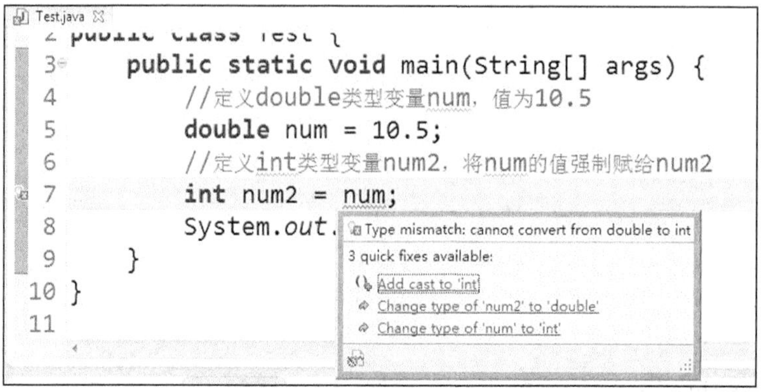

图 2-35

2.5 运算符

2.5.1 赋值运算符

赋值运算符顾名思义就是用来做赋值操作的,更准确地来讲就是将数值赋给某个变量,或者将一个变量的值赋给另外一个变量。语法"数据类型 变量名 = 数值/变量;"表示将等号右边的值赋给等号左边,如代码 2-15 所示。

代码 2-15

```
public class Test {
  public static void main(String[] args) {
    int num = 10;
    int num2 = num;
  }
}
```

例子:

(1)用户张三的体重是 60.5kg,用户李四的体重与张三相同,请用程序描述这一场景。

分析:定义变量 weight1=60.5 表示张三的体重,李四的体重与张三相同,定义李四的体重为 weight2,将 weight1 的值直接赋给 weight2 即可,如代码 2-16 所示。

代码 2-16

```
public class Test {
  public static void main(String[] args) {
    double weight1 = 60.5;
    double weight2 = weight1;
```

```
        System.out.println("李四的体重是"+weight2);
    }
}
```

（2）用户张三的体重为 60.5kg，用户李四的体重为 70.5kg，管理员在录入用户信息时把张三和李四的体重信息搞混了，现要求用程序实现交换张三和李四体重信息的功能。

分析：首先定义两个变量分别表示错误的张三和李四的体重信息，weight1=70.5kg，weight2=60.5kg，然后使用赋值运算符实现 weight1 和 weight2 值的互换。要完成两个数值的交换，不可直接使用 weight1=weigth2 或 weight2=weight1，这样的结果是 weight1 和 weight2 的值相等，要么都等于 60.5，要么都等于 70.5，并没有完成互换。正确的做法是定义第三个变量 weight3，用 weight3 来暂时保存 weight1 的值，然后将 weight2 的值赋给 weight1，最后将 weight3 的值赋给 weight2，如代码 2-17 所示。

代码 2-17

```java
public class Test {
    public static void main(String[] args) {
        double weight1 = 70.5;
        double weight2 = 60.5;
        System.out.println("交换之前张三的体重为："+weight1+",李四的体重为："+weight2);
        double weight3 = weight1;
        weight1 = weight2;
        weight2 = weight3;
        System.out.println("交换之后张三的体重为："+weight1+",李四的体重为："+weight2);
    }
}
```

运行结果如图 2-36 所示。

图 2-36

2.5.2 基本算术运算符

使用基本算术运算符可以完成 Java 程序的基本数学运算，这只适用于数值类型的变量（+除外，它还可以用作字符串拼接）包括+、-、*、/、%、++、--。其中+、-、*、/、%会自动完成操作数的数据类型转换，由低字节转为高字节，比如"10.1+5"，会先将 5 转为 5.0，然后计算 10.1+5.0。这些基本算术运算符的使用方式如下所示。

- 变量 A+ 变量 B：求出变量 A 和变量 B 相加之和。

- 变量 A - 变量 B：求出变量 A 和变量 B 相减之差。
- 变量 A * 变量 B：求出变量 A 和变量 B 相乘之积。
- 变量 A / 变量 B：求出变量 A 和变量 B 相除之商。若变量 A 和变量 B 都为整型，则除不尽时只取商的整数部分，如 10/3 的值为 3。若变量 A 或变量 B 至少有一个是浮点型，则取完整的商，如 10/3.0 的值为 3.3333333333333335（浮点型精度问题导致最后一位为5）。
- 变量 A % 变量 B：求出变量 A 和变量 B 相除之后的余数。
- ++、--的基本语法：变量 A++，++变量 A，变量 A--，--变量 A。
    - ++：求出变量 A+1 的值，等于变量 A+1。
    - --：求出变量 A-1 的值，等于变量 A-1。

注意：变量 A++和++变量 A 是有区别的，变量 A++表示当前操作先取出变量 A 的值，再进行运算。++变量 A 表示当前操作先进行运算，再取出变量 A 的值，如代码 2-18 所示。而--运算也是同理。

代码 2-18

```java
public class Test {
    public static void main(String[] args) {
        int num = 10;
        System.out.println(num++);
    }
}
```

运行结果如图 2-37 所示。

```
@ Javadoc  Declaration  Console ⊠  Progress  Debug
<terminated> Test (141) [Java Application] C:\Program Files (x86)\
10
```

图 2-37

你会发现结果怎么是 10，没有进行+1 运算吗？这是因为我们上面说的在当前操作"System.out.println(num++);"中，会先取出 num 的值进行打印，然后进行+1 运算，所以当前打印的是+1 之前的结果 10，如果我们对代码进行修改，如代码 2-19 所示。

代码 2-19

```java
public class Test {
    public static void main(String[] args) {
        int num = 10;
        System.out.println(num++);
        System.out.println(num);
    }
}
```

运行结果如图 2-38 所示。

图 2-38

如果++放在变量前面呢？如代码 2-20 所示。

代码 2-20

```
public class Test {
    public static void main(String[] args) {
        int num = 10;
        System.out.println(++num);
    }
}
```

运行结果如图 2-39 所示。

图 2-39

先进行+1 运算，再取出 num 的值打印输出，所以是 11。

例子：用程序求出 326 的各位数之和。

分析：本题可分两步来实现，第一步：分解 326，获取各位上的数字；第二步：求出这 3 个数字之和。如代码 2-21 所示。

代码 2-21

```
public class Test {
    public static void main(String[] args) {
        int num = 326;
        int ones = num/100;
        System.out.println("分解 326:");
        System.out.println("百位数是"+ones);
        int tens = num%100/10;
        System.out.println("十位数是"+tens);
        int hundreds = num%10;
        System.out.println("个位数是"+hundreds);
```

```
        int sum = ones+tens+hundreds;
        System.out.println("326 各位数字之和是"+sum);
    }
}
```

运行结果如图 2-40 所示。

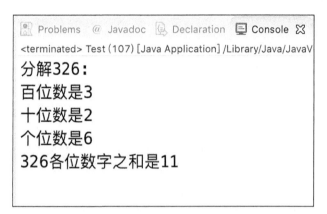

图 2-40

2.5.3 复合算术运算符

复合算术运算符可以在基本算术运算符的基础上进一步简化代码,包括+=、-=、*=、/=、%=。具体的使用方式如下所示。

- 变量 A += 变量 B：先求出变量 A 与变量 B 相加之和,再把计算结果赋给变量 A,等同于变量 A=变量 A+变量 B。
- 变量 A -= 变量 B：等同于变量 A= 变量 A- 变量 B。
- 变量 A *= 变量 B：等同于变量 A= 变量 A* 变量 B。
- 变量 A /= 变量 B：等同于变量 A= 变量 A/ 变量 B。
- 变量 A %= 变量 B：等同于变量 A= 变量 A% 变量 B。

如代码 2-22 所示。

代码 2-22

```
public class Test {
    public static void main(String[] args) {
        int num1 = 9;
        int num2 = 10;
        num1 += num2;
        System.out.println(num1);
        //其他复合算术运算符的使用同上
    }
}
```

运行结果如图 2-41 所示。

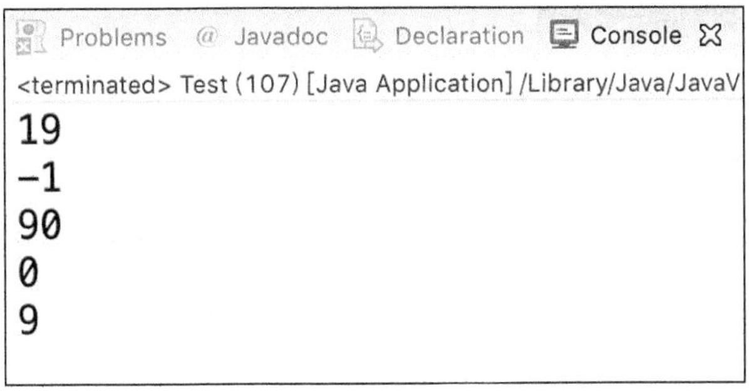

图 2-41

2.5.4 关系运算符

关系运算符用来计算一个表达式是否成立，只有两种结果，即成立或不成立，Java 语言用什么样的数据类型来表示成立或者不成立呢？就是我们之前介绍过的 boolean，boolean 是 Java 的 8 种基本数据类型之一，它的值只有 true 和 false，其中 true 表示成立（真），false 表示不成立（假），实际开发中我们用 boolean 类型的值来表示关系运算符的运算结果。

关系运算符：==、!=、>、<、>=、<=。其中>、<、>=、<=只能用作数值类型的比较，==和!=除了可以比较数值类型，也可以比较两个对象是否相等，对象是 Java 非常重要的一个核心概念。Java 是一门面向对象的编程语言，那什么是面向对象编程？简单理解就是将程序中的所有模块全部抽象成一个个对象，通过这些对象之间的相互调用来完成系统的功能。这部分内容我们会在后续章节详细讲解，这里大家只需要有一个基本认识就可以了。

对象是存储在内存中的，比较两个对象是否相等，实际上就是比较两个对象的内存地址是否相等。如果两个对象的内存地址相等，那就表明它们是同一个对象。==和!=可以分别用来判断两个对象是否相等，两个对象是否不相等，关系运算符的具体使用如下所示。

- 变量 A == 变量 B：变量 A 和变量 B 是否相等，相等的运算结果为 true，不相等的运算结果为 false。

- 变量 A != 变量 B：变量 A 和变量 B 是否不相等，不相等的运算结果为 true，相等的运算结果为 false。

- 变量 A > 变量 B：变量 A 是否大于变量 B，大于的运算结果为 true，等于或小于的运算结果为 false。

- 变量 A < 变量 B：变量 A 是否小于变量 B，小于的运算结果为 treu，等于或大于的运算结果为 false。

- 变量 A >= 变量 B：变量 A 是否大于等于变量 B，大于或等于的运算结果为 true，小于的运算结果为 false。

- 变量 A <= 变量 B：变量 A 是否小于等于变量 B，小于或等于的运算结果为 true，

大于的运算结果为 false。

例子：定义两个 int 类型的变量，值分别为 10 和 11，使用关系运算符判断两个变量是否相等，并用 boolean 类型的变量来接收结果，打印输出，如代码 2-23 所示。

代码 2-23

```java
public class Test {
    public static void main(String[] args) {
        int num1 = 10;
        int num2 = 11;
        boolean result = num1 == num2;
        System.out.println(result);
    }
}
```

运行结果如图 2-42 所示。

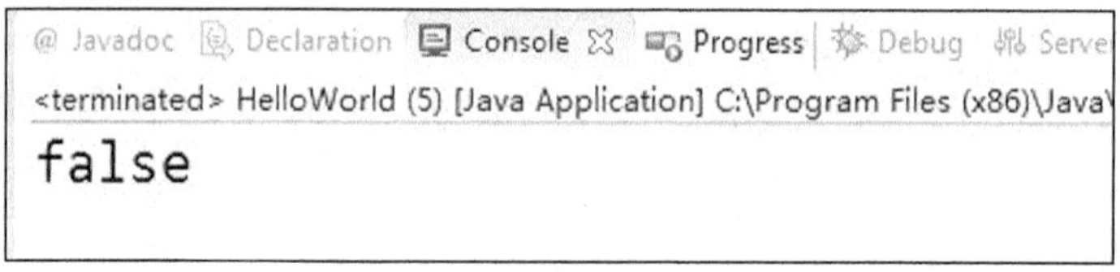

图 2-42

分析一下关键代码 "boolean result = num1 == num2;"，这行代码可以分解成两部分，如图 2-43 所示。

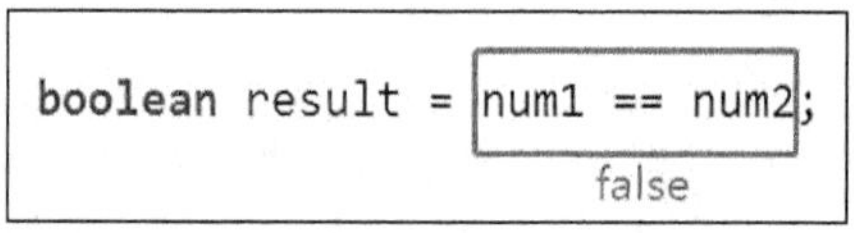

图 2-43

（1）boolean result;   //定义了一个 boolean 类型的变量 result。

（2）result = num1 == num2;   //将 num1 == num2 的值赋给 result。

num1 的值是 10，num2 的值是 11，很显然 num1 不等于 num2，所以 num1 == num2 的值为 false，result 的值就为 false，关系运算符的具体使用如代码 2-24 所示。

代码 2-24

```java
public class Test {
    public static void main(String[] args) {
        int num1 = 10;
        int num2 = 11;
        System.out.println(num1 == num2);
        //其他关系运算符的使用同上
    }
}
```

运行结果如图 2-44 所示。

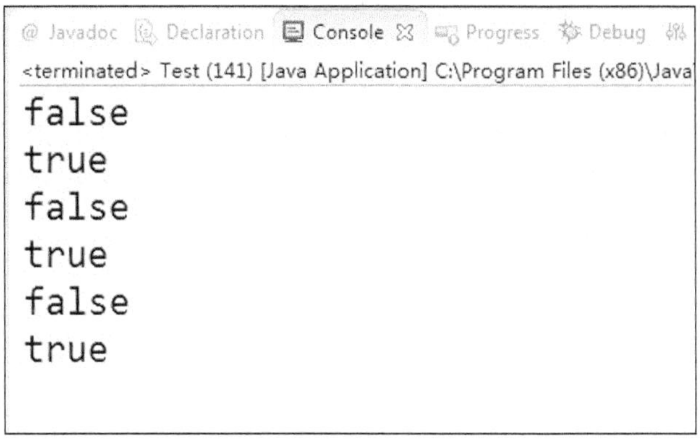

```
false
true
false
true
false
true
```

图 2-44

2.5.5 逻辑运算符

逻辑运算符只能用于 boolean 类型的数据运算，判断 boolean 数据之间的逻辑关系，包括与、或、非这 3 种关系。运算符有&（与）、|（或）、&&（短路与）、||（短路或）、!（非），具体的使用方式如下所示。

- 变量 A & 变量 B：只有当变量 A 和变量 B 都为 true，则结果为 true，否则为 false。
- 变量 A | 变量 B：当变量 A 或变量 B 有一个为 true，结果为 true，否则为 false。
- 变量 A && 变量 B：只有当变量 A 和变量 B 都为 true，结果为 true，否则为 false。
- 变量 A || 变量 B：当变量 A 或变量 B 有一个为 true，则结果为 true，否则为 false。
- !变量 A：若变量 A 为 ture，结果为 false；若变量 A 为 false，结果为 true。

看到这里，小伙伴可能会有些疑问，&和&&，|和||没有区别吗？当然是有区别的，从运算结果的角度讲没有区别，但是运算效率不同，&&和||的效率更高，为什么呢？

与（&）运算中两个操作数必须同时为 true，结果才为 true，也就是说只要有一个操作数为 false，结果肯定是 false。对于 flag1 与 flag2 的与运算，如果 flag1 为 false，则无论 flag2 的值是 true 还是 false，整个运算结果都为 false。所以当我们使用 flag1&&flag2 进行运算时，如果 flag1 的值为 false，就不会去判断 flag2 的值，程序会少执行一次判断，只有当 flag1 为 true 时，才会判断 flag2 的值。

若使用 flag1&flag2 进行运算，无论 flag1 的值为 true 还是 false，程序都会取判断 flag2 的值，所以从效率的角度讲，&&比&更高效。|（或）和||（短路或）同理，如果使用 flag1||flag2，flag1 为 true，则不用判断 flag2 的值，结果肯定为 true，flag1 为 false，才会判断 flag2 的值。而使用 flag1|flag2，无论 flag1 的值是 ture 还是 false，都会去判断 flag2 的值。

例如，有程序如下：

```
int num1 = 10;
int num2 = 11;
System.out.println((num1++==num2)&(++num1==num2));
System.out.println(num1);
```

请给出程序的运行结果。

分析：此题考查的是我们对于算术运算符，关系运算符、逻辑运算符这3种运算符的综合使用能力，第一行打印的是(num1++==num2)&(++num1==num2)，首先来分解这个表达式，运算顺序是：先求出 num1++==num2 的结果，再求出++num1==num2 的结果，然后两个结果进行&运算，如图 2-45 所示。

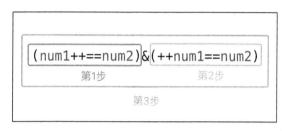

图 2-45

第 1 步：首先进行 num1++的运算，++在操作数之后，所以先使用操作数再进行++运算，此时的num1 为 10，再进行 10==num2 的运算，结果为 false，运算完成之后 num1 的值变为 11。

第 2 步：首先进行++num1 的运算，++在操作数之前，所以先进行++运算再使用操作数，结果为 12，再进行 12==num2 的运算，结果为 false。

第 3 步：false&false 的结果为 false。

所以"System.out.println((num1++==num2)&(++num1==num2));"的结果为 false，"System.out.println(num1);"的结果为 12，如图 2-46 所示。

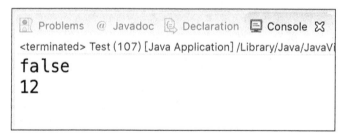

图 2-46

例如，有程序如下：

```
int num1 = 10;
int num2 = 11;
System.out.println((num1++==num2)&&(++num1==num2));
System.out.println(num1);
```

请给出程序的运行结果。

分析：此题和上一题唯一的区别是对两个表达式进行&&运算，结果就完全不同了，我们还是把代码进行分解。首先进行 num1++的运算,此时的 num1 为 10,再进行 10==num2 的运算，结果为 false。表达式 1 的值为 false，无论表达式 2 的值为 true 还是 false，整个运算结果都为 false,这里使用的是&&(短路与),所以此时不会执行表达式 2 的代码,直接返回结果 false,num1 的值为 11,结果如图 2-47 所示。

图 2-47

如果修改代码,将两个表达式的位置互换:

```
int num1 = 10;
int num2 = 11;
System.out.println((++num1==num2)&&(num1++==num2));
System.out.println(num1);
```

首先进行++num1 的运算,结果为 11,再进行 11==num2 的运算,结果为 true。此时还不能判断整个表达式的结果,最终的结果取决于 num1++==num2。继续进行 num1++的运算,此时的 num1 为 11,再进行 11==num2 的运算,结果为 true,true&&true 的结果为 true,同时 num1 进行了两次++运算,结果为 12,如图 2-48 所示。

图 2-48

|(或)和||(短路或)的区别同理。!(非)运算符非常简单,表示对当前操作数进行取反,如代码 2-25 所示。

代码 2-25

```
public class Test {
    public static void main(String[] args) {
        int num1 = 10;
        int num2 = 11;
        boolean flag = num1 == num2;
        boolean flag2 = num1 != num2;
        System.out.println(!flag);
        System.out.println(!flag2);
    }
}
```

运行结果如图 2-49 所示。

图 2-49

2.5.6　条件运算符

条件运算符也叫作三元运算符,可以完成给变量赋值的操作,具体来讲是根据不同的条件给同一个变量赋不同的值,基本语法:变量 A = 条件?值 1:值 2,如果条件成立,值 1 赋给变量 A,否则值 2 赋给变量 A,如代码 2-26 所示。

代码 2-26

```
public class Test {
    public static void main(String[] args) {
        String str = 11>10?"11>10 成立":"11>10 不成立";
        System.out.println(str);
        String str2 = 11<10?"11<10 成立":"11<10 不成立";
        System.out.println(str2);
    }
}
```

运行结果如图 2-50 所示。

图 2-50

例子:商场举办促销活动,会员可参与有奖答题活动,得分大于 80 分即可获得满 100 减 20 优惠券一张,请用程序实现这一场景。

分析:使用条件运算符判断会员得分是否大于 80,以下给出相关信息,如代码 2-27 所示。

代码 2-27

```
public class Test {
    public static void main(String[] args) {
        int score = 60;
        System.out.println("本次答题得分:"+score);
        String result = score > 80?"恭喜您,获得满100减20优惠券一张!":"很遗憾,您没有中奖。";
        System.out.println(result);
    }
}
```

运行结果如图 2-51 所示。

图 2-51

重新答题，再次进行测试，如代码 2-28 所示。

代码 2-28

```
public class Test {
    public static void main(String[] args) {
        int score = 90;
        System.out.println("本次答题得分："+score);
        String result = score > 80?"恭喜您，获得满100减20优惠券一张！":"很遗憾，您没有中奖。";
        System.out.println(result);
    }
}
```

运行结果如图 2-52 所示。

图 2-52

2.5.7 位运算符

位运算符是指对表达式以二进制为单位进行运算，我们知道计算机中的数据全部是以二进制的形式来存储的，而我们日常生活是用十进制来表示数字的。所以我们要学习位运算符的使用，就必须先学会十进制和二进制之间的转换。

十进制转二进制：目标数除以二，若能除尽，该位记做 0，若除不尽，该位记做 1，再对商继续除以二，以此类推，直到商为 0。然后把每一位的结果反序组合就是对应的二进制表示，例如，10 转为二进制就是 1010，转换过程如图 2-53 所示。

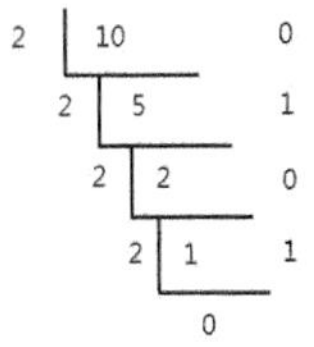

图 2-53

17 转为二进制是 10001，转换过程如图 2-54 所示。

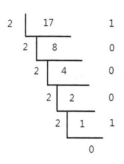

图 2-54

怎么样，是不是非常简单？那么二进制如何转为十进制呢？二级制转十进制，从目标数最右侧算起，本位的数值乘以本位的权重，权重就是 2 的第几位的位数减一次方。如第 1 位就是 2 的（1-1 次）方，也就是 2 的 0 次方，结果为 1；第 2 位就是 2 的（2-1）次方，也就是 2 的 1 次方，结果为 2……以此类推，把每一位数字和本位权重的积相加就是对应的十进制。

例如把二进制的 10001 转为十进制：从最低位算起，第 1 位上的数值为 1，权重是 2 的 1-1 次方，即 2 的 0 次方，所以该位的值为 1 乘以 2 的 0 次方。第 2 位上的数值为 0，权重是 2 的 2-1 次方，即 2 的 1 次方，所以该位的值为 0 乘以 2 的 1 次方，以此类推，如图 2-55 所示。

$$1\ 0\ 0\ 0\ 1$$

$$1*2^{1-1}+\ 0*2^{2-1}+\ 0*2^{3-1}+\ 0*2^{4-1}+\ 1*2^{5-1}$$

$$1*2^0\ +\ 0*2^1\ +\ 0*2^2\ +\ 0*2^3\ +\ 1*2^4$$

$$1\ +\ 0\ +\ 0\ +\ 0\ +\ 16\ =\ 17$$

图 2-55

再来看二进制的 1010 如何转为十进制，如图 2-56 所示。

$$1\ 0\ 1\ 0$$

$$0*2^{1-1}+\ 1*2^{2-1}+\ 0*2^{3-1}+\ 1*2^{4-1}$$

$$0*2^0\ +\ 1*2^1\ +\ 0*2^2\ +\ 1*2^3$$

$$0\ +\ 2\ +\ 0\ +\ 8\ =\ 10$$

图 2-56

学会了十进制和二进制之间的转换，接下来我们来学习如何使用位运算符，位运算符有&（按位与）、|（按位或）、^（按位异或）、<<（左移）、>>（右移）。

我们默认变量 A 和变量 B 是二进制数据，具体的使用方式如下所示。

- 变量 A & 变量 B：每一位的数字一一对应，若都为 1，则该位记做 1，否则为 0。
- 变量 A | 变量 B：每一位的数字一一对应，只要有一个为 1，则该位记做 1，否则为 0。
- 变量 A ^ 变量 B：每一位的数字一一对应，相同记做 0，不同记做 1。
- 变量 A << 变量 B：变量 A 乘以 2 的 B 次方，如 2 << 3，结果为 2*8。
- 变量 A >> 变量 B：变量 A 除以 2 的 B 次方，如 2 >> 3，结果为 2/8。

例子：分别求出 10&5、10|5、10^3、2 << 3、2 >> 3 的值（详见代码 2-29）。

分析：10&5，此时的&不是逻辑运算符，而是位运算符，如何区分逻辑运算符和位运算符？通过操作数来判断，如果操作数是 boolean 类型，则&为逻辑运算符；如果操作数为数值类型，则&为位运算符。第 1 步：将两个操作数 10 和 5 分别转为二进制，10 的二级制是 1010，5 的二级制是 101。第 2 步：对 1010 和 101 进行&（按位与）运算，每一位的数字一一对应，该位没有数字的补 0，如图 2-57 所示。

```
1 0 1 0
0 1 0 1
―――――――
0 0 0 0
```

图 2-57

所以 10&5 的结果是 0000，转为十进制为 0，同理 10|5 运算过程如图 2-58 所示。

```
1 0 1 0
0 1 0 1
―――――――
1 1 1 1
```

图 2-58

10|5 的结果为 1111，转为十进制是 15，10^3 运算过程如图 2-59 所示。

```
1 0 1 0
0 0 1 1
―――――――
1 0 0 1
```

图 2-59

10^3 的结果为 1001，转为十进制是 9。

代码 2-29

```
public class Test {
   public static void main(String[] args) {
      System.out.println(10&5);
      System.out.println(10|5);
      System.out.println(10^3);
      System.out.println(2<<3);
      System.out.println(2>>3);
   }
}
```

运行结果如图 2-60 所示。

```
0
15
9
16
0
```

图 2-60

2.6 小结

本章首先为大家讲解了如何使用 Eclipse 集成开发环境来编写 Java 程序，除了 Eclipse，主流的集成开发环境还有 IntelliJ IDEA，本书使用的是 Eclipse。无论使用哪种开发工具都是为了帮助开发者更加高效、快速地完成企业级项目开发工作，让开发者不再需要关注底层实现，并且有代码自动提示、智能代码助手等功能。本章还讲解了 Java 中的变量、基本数据类型、运算符等最基本的语法，学习完本章内容，大家就可以用 Java 编写一个功能简单的小程序了。

第 3 章 Java 进阶

上一章我们学习了 Java 的基本语法、变量等，相信大家已经可以使用 Java 编写简单的小程序了。本章深入学习 Java 的逻辑处理机制，让我们开发的 Java 程序功能更加完善。

3.1 流程控制

在实际开发中，我们经常会遇到这样的情况，程序在不同的条件下，需要执行不同的逻辑代码。那么如何来实现呢？可以通过流程控制让程序分情况完成不同的业务逻辑，比如 2.5.6 节举过的例子"商场举办促销活动"，可以通过流程控制对代码进行优化。

3.1.1 if-else

if-else 是一个基本的流程控制语法，用于判断某个条件是否成立，然后执行不同的逻辑，基本语法如下：

```
if(判断条件){
//条件成立的代码
}else{
//条件不成立的代码
}
```

例子：商场举办促销活动，会员可参与有奖答题活动，得分大于 80 分即可获得满 100 减 20 优惠券一张，否则没有奖品。请用 if-else 实现这一场景。

分析：这是一个典型的 if-else 逻辑判断，条件是会员答题得分是否大于 80 分，具体实现如代码 3-1 所示。

代码 3-1

```
public class Test {
    public static void main(String[] args) {
        int score = 100;
```

```
        System.out.println("本次答题得分："+score);
        if(score > 80) {
            System.out.println("恭喜您，获得满100减20优惠券一张！");
        }else {
            System.out.println("很遗憾，您没有中奖。");
        }
    }
}
```

运行结果如图 3-1 所示。

图 3-1

将 score 改为 70，如代码 3-2 所示。

代码 3-2

```
public class Test {
    public static void main(String[] args) {
        int score = 70;
        System.out.println("本次答题得分："+score);
        if(score > 80) {
            System.out.println("恭喜您，获得满100减20优惠券一张！");
        }else {
            System.out.println("很遗憾，您没有中奖。");
        }
    }
}
```

运行结果如图 3-2 所示。

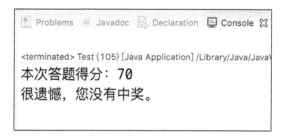

图 3-2

非常简单的逻辑！现在促销活动升级了，如果会员答题得分大于 60，同时积分达到 500 分；或者答题得分大于 80，同时积分达到 300 分，可获得满 300 减 100 优惠券一张，如何实现呢？

分析：此题还是一个 if-else 的逻辑判断，只不过条件变成了答题得分大于 60，同时

积分>=500；或者答题得分大于80，同时积分>=300。多个表达式同时构成条件，并且表达式之间有逻辑关系，我们可以使用关系运算符结合逻辑运算符来完成（答题得分> 60 &&积分>=500）||（答题得分> 80 &&积分>=300），具体实现如代码3-3所示。

代码3-3

```java
public class Test {
    public static void main(String[] args) {
        int score = 90;
        int integral = 600;
        System.out.println("本次答题得分："+score);
        System.out.println("您的会员积分："+integral);
        if((score > 60 && integral >= 500) || (score > 80 && integral >= 300)){
            System.out.println("恭喜您，获得满300减100优惠券一张！");
        }else{
            System.out.println("很遗憾，您没有中奖。");
        }
    }
}
```

运行结果如图3-3所示。

图 3-3

这里需要注意的是在代码语句"(score > 60 && integral >= 500) || (score > 80 && integral >= 300)"中，为什么要加小括号，不加的话会不会有影响？如果不加也不会有影响，现在的逻辑结构是先执行两个&&运算，再执行一个||运算，如图3-4所示。

图 3-4

如果不加小括号，代码为"score > 60 && integral >= 500 || score > 80 && integral >= 300"，逻辑顺序保持不变。在多种运算符混合使用的情况下，程序会根据运算符的优先级来决定不同运算的执行顺序，运算符优先级顺序为！> 算术运算符 > 关系运算符 > && > ||。

3.1.2 多重if

if-else是构成一个逻辑分支的基本结构，描述了程序可能执行的两种逻辑。如果一个程序可能会存在 3 种甚至更多种逻辑，我们应该如何实现呢？这种情况下可以使用多重 if。

例子：买衣服的时候根据身高来选择对应的尺码，身高（单位：cm）173 以下为 M 码，173~178 为 L 码，178 以上为 XL 码，请用程序实现这一逻辑。

分析：本题需要我们根据身高划分出 3 个逻辑块，分别是身高大于 178，身高（单位：cm）大于等于 173 并且小于等于 178，身高小于 173，然后分别打印不同的信息，具体实现如代码 3-4 所示。

代码 3-4

```java
public class Test {
    public static void main(String[] args) {
        int height = 176;
        if(height > 178){
            System.out.println("XL 码");
        }
        if(height >= 173 && height <= 178){
            System.out.println("L 码");
        }
        if(height < 173){
            System.out.println("M 码");
        }
    }
}
```

运行结果如图 3-5 所示。

图 3-5

我们对代码进行修改，使用 if-else 的结构来完成，if-else 首先需要划分出两个逻辑块，具体思路应该是这样的：先划分出身高（身高：cm）大于 178 和身高小于等于 178，再将身高小于等于 178 的代码块划分为身高大于等于 173 和身高小于 173，具体实现如代码 3-5 所示。

代码 3-5

```java
public class Test {
    public static void main(String[] args) {
        int height = 176;
        if(height > 178){
            System.out.println("XL 码");
        }else {
            if(height >= 173) {
                System.out.println("L 码");
            }else {
                System.out.println("M 码");
```

            }
         }
      }
   }
```

同时还可以再次变换写法，如代码 3-6 所示。

代码 3-6

```
public class Test {
   public static void main(String[] args) {
      int height = 176;
      if(height > 178){
         System.out.println("XL 码");
      }else if(height >= 173) {
         System.out.println("L 码");
      }else {
         System.out.println("M 码");
      }
   }
}
```

if-else 的使用是比较灵活的，可以根据不同情况选择不同的组合方式，需要注意的是：

- if 后面必须跟（条件）；
- else 后面不能跟（条件）；
- else 后面可直接跟{}，也可跟 if 语句。

3.1.3　if 嵌套

我们再次升级难度，如果遇到条件中包含子条件，那么使用单一的 if-else 或者多重 if 已经不能满足需求了。因为单一的 if-else 和多重 if 只能在一个维度处理逻辑，如果再增加一个维度，就需要使用 if 嵌套来完成。

例子：某组织举办电子竞技大赛，预赛成绩在 80 分以上的可以进入决赛，根据年龄分别进入 A 组（20 岁及 20 岁以上）和 B 组（18～19 岁），请用程序实现这一场景。

分析：本题是一个典型的条件中包含子条件的逻辑结构，第一维的条件是成绩大于 80 分，在此条件成立的基础上，第二维的条件是年龄大于等于 20 岁和年龄大于等于 18 并且小于 20，所以使用 if 嵌套可以实现，如代码 3-7 所示。

代码 3-7

```
public class Test {
   public static void main(String[] args) {
      int score = 90;
      int age = 22;
      if(score > 80){
         if(age >= 20){
            System.out.println("成功晋级 A 组");
         }else if(age >= 18){
```
```

```
            System.out.println("成功晋级B组");
        }
      }
    }
}
```

运行结果如图 3-6 所示。

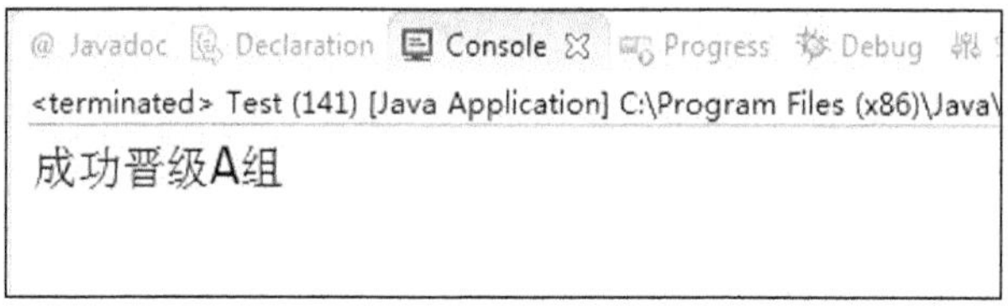

```
@ Javadoc  Declaration  Console ⊠  Progress  Debug
<terminated> Test (141) [Java Application] C:\Program Files (x86)\Java\
成功晋级A组
```

图 3-6

3.1.4　switch-case

switch-case 也可以完成流程控制，与 if 不同的是，switch-case 只能完成等值判断，即条件如果是判断两个值是否相等，可以使用 switch-case，如果是比较两个值的大小关系，则不能使用 switch-case。switch 支持 int、short、byte、char、枚举、String 数据类型的判断，不支持 boolean 类型。

```
基本语法：switch(变量){
case 值1:
代码1；
break;
case 值2:
代码2；
break;
……
default:
代码n；
break;
}
用if-else来类比，等同于
if(变量==值1){
代码1；
}
else if(变量==值2){
代码2；
}
……
else{
代码n；
}
```

case 判断变量是否等于某个值，default 表示所有 case 都不成立的情况下所执行的代码。

例子：小明参加马拉松比赛，获得第 1 名奖励 2000 元，获得第 2 名奖励 1000 元，获得第 3 名奖励 500 元，否则没有奖励。请分别用 if-else 和 switch-case 实现这一场景。

分析：本题考察的是 if-else 和 switch-case 的使用，非常简单的逻辑判断，具体实现如代码 3-8 所示。

代码 3-8

```java
public class Test {
    public static void main(String[] args) {
        int placing = 1;
        if(placing == 1){
            System.out.println("奖励2000元");
        }else if(placing == 2){
            System.out.println("奖励1000元");
        }else if(placing == 3){
            System.out.println("奖励500元");
        }else{
            System.out.println("没有奖励");
        }

        switch (placing) {
        case 1:
            System.out.println("奖励2000元");
            break;
        case 2:
            System.out.println("奖励1000元");
            break;
        case 3:
            System.out.println("奖励500元");
            break;
        default:
            System.out.println("没有奖励");
            break;
        }
    }
}
```

运行结果如图 3-7 所示。

图 3-7

需要注意的是，每一个 case 语句必须要跟 break，表示结束当前的代码块，如果不跟 break，则从当前的 case 语句起，后面所有的 case 判断都失效，会直接执行对应的代码块，如代码 3-9 所示。

代码 3-9

```java
public class Test {
    public static void main(String[] args) {
        int placing = 1;
```

```
        switch (placing) {
        case 1:
            System.out.println("奖励 2000 元");
        case 2:
            System.out.println("奖励 1000 元");
        case 3:
            System.out.println("奖励 500 元");
        default:
            System.out.println("没有奖励");
        }
    }
}
```

运行结果如图 3-8 所示。

图 3-8

可以看到，case 1 是成立的，但是因为没有加 break，所以后续的代码块全部执行而且不再进行判断。

3.2 循环

循环是程序开发非常重要，也是使用频率很高的一个技能点，为什么要使用循环？举一个例子，要求将 "Hello World" 打印 10 遍。如果不使用循环，需要写 10 行相同的代码 "System.out.println("Hello World");"，如代码 3-10 所示。

代码 3-10

```
public class Test {
    public static void main(String[] args) {
        System.out.println("Hello World");        //该语句写 10 次
    }
}
```

很显然这种写法并不是最优的方式，重复性的代码非常多。如果使用循环，我们就可以避免这个缺陷，用效率更高的方式来完成同样的需求。

3.2.1 while 循环

while 是一种具体的循环方式，它的基本语法如下：

```
while(循环条件){
    //循环体
}
```

当循环条件成立时，会重复执行循环体中的代码，直到条件不成立。那么很显然，我们应该在循环体中动态控制循环条件是否成立，否则就形成了死循环。循环条件始终成立导致循环体一直执行，不会停止。所以我们在写循环时，需要额外添加控制循环条件的代码，这部分代码和循环本身组成了循环四要素，这是很重要的知识点，它描述了循环的本质，循环四要素如下：

- 初始化循环变量；
- 循环条件；
- 循环体；
- 更新循环变量。

一段完整的循环代码，四要素缺一不可。接下来我们使用 while 循环来实现重复打印 10 次"Hello World"的需求，如代码 3-11 所示。

代码 3-11

```java
public class Test {
    public static void main(String[] args) {
        int num = 0;
        while(num < 10){
            System.out.println("Hello World");
            num++;
        }
    }
}
```

运行结果如图 3-9 所示。

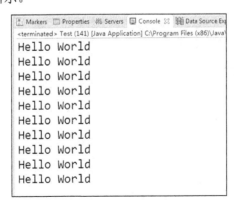

图 3-9

对上述代码进行分析，循环四要素如图 3-10 所示。

执行顺序：第一步，初始化循环变量。第二步，判断循环条件，若成立，则执行循环体；若不成立，则直接跳过循环，去执行后面的代码。若执行了循环体，那第三步就更新循环变量。然后再又回到第二步，重新判断循环条件是否成立，如图 3-11 所示。

图 3-10

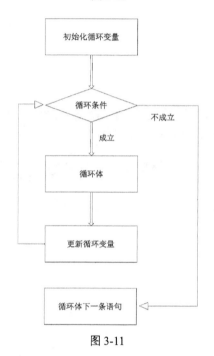

图 3-11

例子：使用 while 循环完成一个学生成绩查询系统，输入学生编号，查询出该学生的成绩，输入"yes"可重复查询，输入"no"结束循环，如图 3-12 所示。

图 3-12

具体实现如代码 3-12 所示。

代码 3-12

```java
import java.util.Scanner;
public class Test {
    public static void main(String[] args) {
        System.out.println("学生成绩查询系统");
        Scanner scanner = new Scanner(System.in);
        int num;
        String str = "yes";
        while(str.equals("yes")){
            System.out.print("请输入学生编号：");
            num = scanner.nextInt();
            switch(num){
                case 1:
                    System.out.println("张三的成绩为90");
                    break;
                case 2:
                    System.out.println("李四的成绩为96");
                    break;
                case 3:
                    System.out.println("王五的成绩为88");
                    break;
            }
            System.out.print("是否继续?yes/no：");
            str = scanner.next();
        }
        System.out.println("查询结束");
    }
}
```

Scanner 是一个工具类，用来接收用户在控制台输入的数据，使用 Scanner 首先需要将 Scanner 类引入当前类中，语句："import java.util.Scanner;"，其中 java.util 是 Scanner 类的包名，后面我们会详细讲解什么是包，这里需要注意在一个类中要使用其他包中的类，必须要引入（即 import）。引入之后来创建该类的实例化对象"Scanner scanner = new Scanner(System.in);"，其中"System.in"表示获取的是输入数据。接收不同的数据需要调用不同的方法，nextInt()方法用来获取 int 类型的数据，next()方法用来获取 String 类型的数据。判断 String 类型的数据是否相等时，不能使用==，需要使用 equals()方法进行判断。

3.2.2　do-while 循环

do-while 循环和 while 循环很类似，区别在于 do-while 循环会先执行一次循环体，再做判断，而 while 循环必须先判断，再决定是否执行循环体。如果循环条件不成立，do-while 循环会执行一次循环体，while 循环一次都不会执行，基本语法如下。

```
do{
//循环体
}while(循环条件)
```

do-while 同样遵守循环四要素的原则，使用 do-while 循环来实现重复打印 10 次"Hello World"的需求，如代码 3-13 所示。

代码 3-13

```java
public class Test {
    public static void main(String[] args) {
        int num = 0;
        do{
            System.out.println("Hello World");
            num++;
        }while(num < 10);
    }
}
```

对上述代码进行分析，循环四要素如图 3-13 所示。

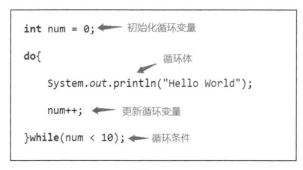

图 3-13

do-while 循环是先执行后判断，顺序如图 3-14 所示。

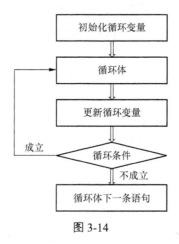

图 3-14

例子：张三参加体能测试 1000 米跑，如果成绩不合格，则继续测试，直到合格为止，使用循环来实现这一场景。

分析：本题是一个典型的循环代码，如果不合格，则需要循环测试。同时需要注意的是，这里应该先执行一次测试，再来判断是否合格，所以使用 do-while 循环最合适，先执行后判断，具体实现如代码 3-14 所示。

代码 3-14

```java
import java.util.Scanner;
public class Test {
    public static void main(String[] args) {
        Scanner scanner = new Scanner(System.in);
        String str;
        do{
            System.out.println("张三参加体能测试，跑1000米...");
            System.out.print("是否合格?yes/no：");
            str = scanner.next();
        }while(str.equals("no"));
        System.out.println("合格，通过测试！");
    }
}
```

运行结果如图 3-15 所示。

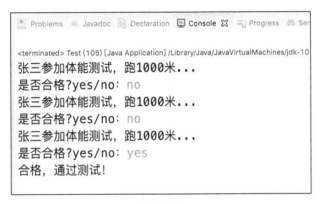

图 3-15

3.2.3 for 循环

for 循环也是很常用的一种循环语法，它与 while 循环的最大区别在于 while 循环适用于循环次数不确定的场景，如张三参加体能测试，成绩合格就停止循环，否则循环一直进行，这时循环次数是不确定的。for 循环适用于循环次数确定的场景，如输出 10 次 "Hello World"，很明确要执行 10 次循环，基本语法：

```
for(初始化循环变量;循环条件;更新循环变量){
//循环体
}
```

使用 for 循环来实现重复打印 10 次 "Hello World" 的需求，如代码 3-15 所示。

代码 3-15

```java
public class Test {
    public static void main(String[] args) {
        for(int i = 0; i < 10; i++){
            System.out.println("Hello World");
        }
```

        }
    }

对上述代码进行分析，循环四要素如图 3-16 所示。

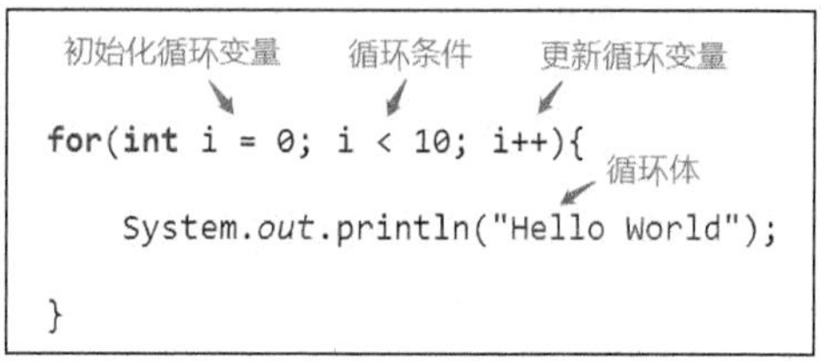

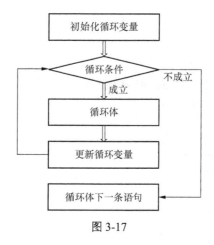

图 3-16

for 循环是先判断，后执行，流程如图 3-17 所示。

图 3-17

3.2.4　while、do-while 和 for 这 3 种循环的区别

while、do-while 和 for 这 3 种循环结构是开发中经常会使用到的技能点，三者有很多相同点，同时也有很多差异。它们各有各的特点，我们需要根据具体的业务场景来选择最合适的循环方式。

相同点：都遵循循环四要素，即初始化循环变量、循环条件、循环体、更新循环变量。

不同点：

- while 和 do-while 适用于循环次数不确定的场景；for 适用于循环次数确定的场景。
- while 和 for 是先判断循环条件，再执行循环体；do-while 是先执行循环体，再判断循环条件。

例子：分别使用 while、do-while、for 循环输出 10 以内的所有奇数，具体实现如代码 3-16 所示，"System.out.print();" 表示不换行输出。

代码 3-16

```
public class Test {
```

```java
    public static void main(String[] args) {
        //while 循环
        int num = 0;
        while(num <= 10){
            if(num%2!=0) {
                System.out.print(num+",");
            }
            num++;
        }
        System.out.println("");
        //do-while 循环
        int num2 = 0;
        do{
            if(num2%2!=0) {
                System.out.print(num2+",");
            }
            num2++;
        }while(num2 <= 10);
        System.out.println("");
        //for 循环
        for(int i = 1; i <= 10;i++){
            if(i%2!=0) {
                System.out.print(i+",");
            }
        }
    }
}
```

运行结果如图 3-18 所示。

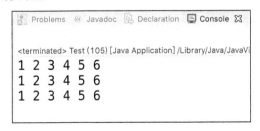

图 3-18

3.2.5 双重循环

上一节我们讲解的 while、do-while、for 都是一维结构的循环，即在一个维度重复执行代码，这种方式的循环不能完成较为复杂的需求（比如重复打印 3 行 "1 2 3 4 5 6"），如图 3-19 所示，该如何实现呢？

图 3-19

分析：要求打印一个 3×6 的二维数字信息，共 3 行，每行 6 个数字。问题可分解为两步：第一步，打印 6 个数字；第二步，将第一步操作循环执行 3 次。我们首先使用一维循环来处理，循环次数确定，所以选择 for 循环，一次循环打印 6 个数字，那么只要执行 3 次同样的 for 循环即可实现需求，具体实现如代码 3-17 所示。

代码 3-17

```java
public class Test {
    public static void main(String[] args) {
        for(int i = 1;i <= 6;i++){
            System.out.print(i+" ");
        }
        //换行
        System.out.println("");
        //重复上述代码 2 次
    }
}
```

我们分析一下上述程序，会发现 3 次 for 循环的代码是完全相同的，如图 3-20 所示。当程序中重复出现完全相同的代码时，就需要考虑用循环来进行优化。

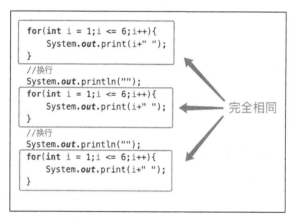

图 3-20

具体的优化思路是把 for 循环作为一个循环体，放入另外一个循环中，如代码 3-18 所示。

代码 3-18

```java
public class Test {
    public static void main(String[] args) {
        //外层循环控制打印 3 行
        for(int j = 0;j < 3;j++){
            //内存循环控制每行打印 6 个数字
            for(int i = 1;i <= 6;i++){
                System.out.print(i+" ");
            }
            System.out.println("");
        }
    }
}
```

现在难度升级,请打印出如图 3-21 所示的图案。

```
 Problems  @ Javadoc  Declaration  Console ⊠

<terminated> Test (105) [Java Application] /Library/Java/JavaV
       0
      1 2
     2 3 4
    3 4 5 6
     2 3 4
      1 2
       0
```

图 3-21

分析:此图较为复杂,上下左右都是对称的。先拆分上下部分,考虑如何实现上半部分,第 1 行先打印 3 个空格,再打印 1 个数字;第 2 行先打印 2 个空格,再打印 2 个数字;第 3 行先打印 1 个空格,再打印 3 个数字;第 4 行先打印 0 个空格,再打印 4 个数字。由此找出规律:我们规定行数从 0 开始,每行的空格数=3-行数,每行的数字个数=行数+1。按照这个思路,具体实现如代码 3-19 所示。

代码 3-19

```java
public class Test {
    public static void main(String[] args) {
        for(int j = 0; j < 4; j++){
            for(int k = 0;k < 3-j;k++){
                System.out.print(" ");
            }
            for(int i = j; i < 2*j+1; i++){
                System.out.print(i+" ");
            }
            System.out.println("");
        }
    }
}
```

运行结果如图 3-22 所示。

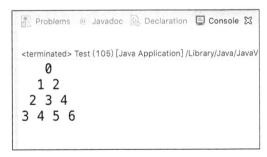

图 3-22

上半部分成功打印之后,按照同样的思路打印出下半部分即可,第 1 行先打印 1 个空格，再打印 3 个数字；第 2 行先打印 2 个空格，再打印 2 个数字；第 3 行先打印 3 个空格，再打印 1 个数字。由此找出规律：我们规定行数从 0 开始，每行的空格数=行数+1，每行的数字个数=3−行数。按照这个思路，具体实现如代码 3-20 所示。

代码 3-20

```java
public class Test {
    public static void main(String[] args) {
        for(int j = 0; j < 4; j++){
            for(int k = 0;k < 3-j;k++){
                System.out.print(" ");
            }
            for(int i = j; i < 2*j+1; i++){
                System.out.print(i+" ");
            }
            System.out.println("");
        }
        for(int j = 0; j < 3; j++){
            for(int k = 0;k < j+1;k++){
                System.out.print(" ");
            }
            //打印对应的数字，i 从 2-j 开始，i < 5-2*j
            for(int i = 2-j; i < 5-2*j; i++){
                System.out.print(i+" ");
            }
            System.out.println("");
        }
    }
}
```

例子：用双重循环打印九九乘法口诀表。具体实现如代码 3-21 所示。

代码 3-21

```java
public class Test {
    public static void main(String[] args) {
        for(int i = 1;i<=9;i++){
            for(int j = 1; j <= i; j++){
                System.out.print(i+"*"+j+"="+i*j+"\t");
            }
            System.out.println("");
        }
    }
}
```

运行结果如图 3-23 所示。

```
1*1=1
2*1=2   2*2=4
3*1=3   3*2=6   3*3=9
4*1=4   4*2=8   4*3=12  4*4=16
5*1=5   5*2=10  5*3=15  5*4=20  5*5=25
6*1=6   6*2=12  6*3=18  6*4=24  6*5=30  6*6=36
7*1=7   7*2=14  7*3=21  7*4=28  7*5=35  7*6=42  7*7=49
8*1=8   8*2=16  8*3=24  8*4=32  8*5=40  8*6=48  8*7=56  8*8=64
9*1=9   9*2=18  9*3=27  9*4=36  9*5=45  9*6=54  9*7=63  9*8=72  9*9=81
```

图 3-23

3.2.6 终止循环

通过前面的各节我们了解到只要循环条件成立，循环体就会一直执行。那么如果遇到某种特定情况，需要强制停止循环，应该怎么做呢？手动终止循环有两种方式：break 和 continue。两者的区别是：break 表示跳出整个循环体，continue 表示跳出本次循环，进入下一次循环。两者适用于所有循环结构，都必须写在循环体代码中。

例子：将 1～100 的整数依次相加，当总和大于 50 时跳出循环，并输出结果。具体实现如代码 3-22 所示。

代码 3-22

```java
public class Test {
    public static void main(String[] args) {
        int sum = 0;
        for(int i = 1;i <= 100;i++){
            sum += i;
            if(sum > 50){
                break;
            }
        }
        System.out.println("总和是"+sum);
    }
}
```

运行结果如图 3-24 所示。

图 3-24

当总和大于 50 时，需要结束整个循环体，所以使用 break。

例子：计算 1～200 的所有奇数之和，具体实现如代码 3-23 所示。

代码 3-23

```java
public class Test {
    public static void main(String[] args) {
        int sum = 0;
        for(int i = 1;i <= 200;i++){
            if(i%2 == 0){
                continue;
            }
            sum += i;
        }
        System.out.println("1-200 的奇数和是"+sum);
```

        }
    }

运行结果如图 3-25 所示。

图 3-25

3.3 数组

3.3.1 什么是数组

在学习数组之前我们先来回顾一下之前的内容,计算机将程序中需要用到的数据保存在内存中。直接通过内存地址去取值的方式很麻烦,不利于代码的开发,所以引入了变量的概念,相当于给内存起了一个别名。通过自定义的变量名就可以找到内存中存储数据的位置,进而取出其中的数据。思考这样一个问题,如果是信息量较大的一组数据,用哪种方式保存效率更高呢?

我们当然可以创建多个变量来表示这些数据,但是很显然这种方式并不是很合理,变量个数太多不利于统一管理,并且变量的命名也会有很多限制。就好比我们管理了一间货物随意摆放、没有分门别类进行整理的超市,会是一件很苦恼的事情。程序也一样,我们应该对这种存储方式进行优化,对于超市来说就是将不同种类的货物统一摆放到对应的货架上。程序中也有具有类似功能的一种数据结构,这就是数组。数组就是一种可以存放大量数据类型相同的变量的数据结构,是一个具有相同数据类型的数据集合。数组中的数据必须是同一种数据类型,可以类比现实生活中学校的所有学生、机房里的所有计算机等。我们知道使用变量来保存数据,需要在内存中开辟一块空间,而数组就是在内存中开辟一串连续的空间来保存数据,如图 3-26 所示。

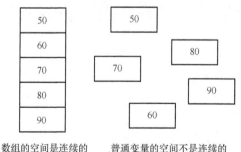

图 3-26

3.3.2 数组的基本要素

了解完数组的基本概念，接下来我们学习数组的结构。一个数组由 4 种基本元素构成：(1) 数组名称；(2) 数组元素；(3) 元素下标；(4) 数据类型。数组本身也是一个变量，既然是一个变量，就需要定义变量名，也就是数组名称。数组中保存的每一个数据都会有一个下标（从 0 开始），相当于编号，通过编号可以快速检索到对应的元素。数组中的所有元素必须是同一种数据类型，如图 3-27 所示。

例子：下列哪组数据可以构成一个数组？如果能，应该选择哪种数据类型？

(1) "张三"、"李四"、true、"王五"

(2) 10.5、66、"小明"、'男'

(3) 12、33、66、80

(4) 22.5f、33.6f、100.0f、202.2f

图 3-27

第 1 组不可以，因为同时包含了 String 类型和 boolean 类型。第 2 组不可以，因为同时包含了 double 类型、int 类型、String 类型、char 类型。第 3 组可以，因为数组的类型均为 int。第 4 组可以，因为数组的类型均为 float。

3.3.3 如何使用数组

数组也是一个变量，创建数组的步骤和创建普通变量基本一致，具体步骤如下。

(1) 声明数组：数据类型[] 数组名；如 "int[] array1;" 表示声明了一个 int 类型的数组，该数组中只能存放 int 类型的数据，"String[] array2;" 表示声明了一个 String 类型的数组，该数组中只能存放 String 类型的数据。

(2) 分配内存空间：创建数组必须要指定数组的长度，根据指定长度在内存中开辟一串连续的空间，并且长度不能修改。语法："数组名=new 数据类型[数组长度]"。例如 "array1 = new int[6];" 表示 array1 的长度为 6；例如 "array2 = new String[7];" 表示 array2 的长度为 7。

(3) 给数组赋值：分配完内存空间就可以向数组中存值了，通过下标找到数组中对应的内存空间，完成赋值，相当于通过编号找到储物柜，然后把需要保存的物品放入柜子。如 "array1[0] = 1;array1[2] = 3;array2[1] = "Java";array2[3] = "Hello World";"。

(4) 使用数组：当数组创建完成后，可以通过下标获取数组中的数据，相当于通过编号找到储物柜，然后取出柜子中保存的物品。

例子：张三身高（单位：cm）为 179，李四身高为 182，王五身高为 167，小明身高为 176，要求创建一个数组，保存 4 个用户的身高，并且求出平均值，具体实现如代码 3-24 所示。

代码 3-24

```
public class Test {
```

```java
    public static void main(String[] args) {
        int[] array;
        array = new int[4];
        array[0] = 179;
        array[1] = 182;
        array[2] = 167;
        array[3] = 176;
        //遍历数组计算平均值,数组长度可以通过 length 属性获取
        double sum = 0;
        for(int i = 0;i < array.length;i++){
            sum += array[i];
        }
        double avg = sum/array.length;
        System.out.println("平均身高是："+avg);
    }
}
```

运行结果如图 3-28 所示。

图 3-28

上述代码可以进一步简化，声明数组和开辟空间可以用一行代码完成，如代码 3-25 所示。

代码 3-25

```java
public class Test {
    public static void main(String[] args) {
        int[] array = new int[4];
        array[0] = 179;
        array[1] = 182;
        array[2] = 167;
        array[3] = 176;
        //遍历数组计算平均值,数组长度可以通过 length 属性获取
        double sum = 0;
        for(int i = 0;i < array.length;i++){
            sum += array[i];
        }
        double avg = sum/array.length;
        System.out.println("平均身高是："+avg);
    }
}
```

声明数组时，[]可以放在数据类型的后面，也可以放在数组名后面，即 int[] array 和 int array[]都可以。边声明边赋值有两种方式，如代码 3-26 所示。

代码 3-26

```
public class Test {
    public static void main(String[] args) {
        int[] array1 = {10,20,30};
        int[] array2 = new int[]{10,20,30};
    }
}
```

数组与我们之前介绍的变量在内存中的保存方式是不同的。内存可以简单分为栈内存和堆内存。基本数据类型的变量和值都保存在栈内存中，在栈内存开辟的空间中直接放入数据，比如 int num=10 在栈内存中的存储如图 3-29 所示。

引用数据类型的变量保存在栈内存中。而变量的值，也就是引用实际指向的对象，保存在堆内存中，即栈内存保存的是堆内存的地址。什么是引用数据类型？这里大家可以简单记为只要是通过 new 关键字创建的变量都是引用类型，如 String 和数组。int[] array = {50,60,70,80,90}在内存中的存储如图 3-30 所示。

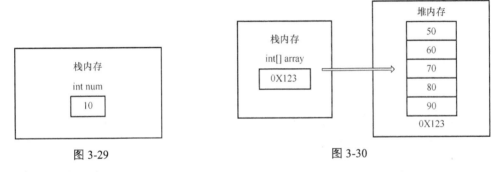

图 3-29　　　　　　　　　　图 3-30

常见错误如下。

（1）数组声明时数据类型不匹配，错误示例如代码 3-27 所示。

代码 3-27

```
public class Test {
    public static void main(String[] args) {
        int[] array = new String[3];
    }
}
```

Eclipse 会提示错误信息，并给出解决方法：将 array 的类型改为 String[]，保证前后一致，如图 3-31 所示。

图 3-31

（2）边声明边赋值必须写在同一行，错误示例如代码 3-28 所示。

代码 3-28

```java
public class Test {
    public static void main(String[] args) {
        int[] array = new int[3];
        array = {10,20,30};
        int[] array2;
        array2 = {10,20,30};
    }
}
```

运行代码，报错信息如图 3-32 所示。

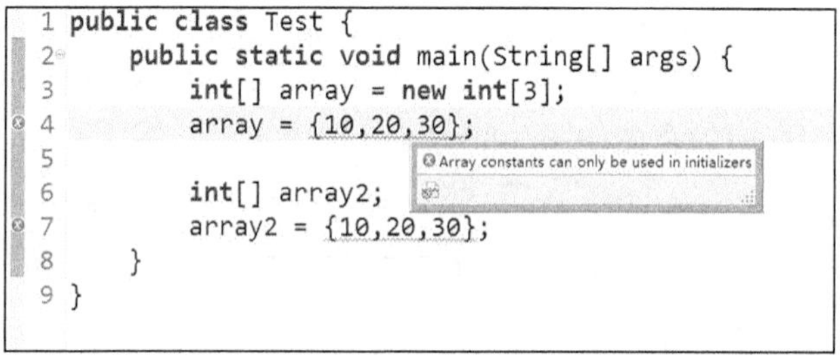

```
1  public class Test {
2      public static void main(String[] args) {
3          int[] array = new int[3];
4          array = {10,20,30};
5                              Array constants can only be used in initializers
6          int[] array2;
7          array2 = {10,20,30};
8      }
9  }
```

图 3-32

（3）数组下标越界，通过下标取值时给出的下标超出了数组的长度范围。这里需要注意的是，数组下标是从 0 开始的，如果数组长度为 3，下标的界限为 0~2，如果数组长度为 5，下标的界限为 0~4，错误示例如代码 3-29 所示。

代码 3-29

```java
public class Test {
    public static void main(String[] args) {
        int[] array = {10,20,30};
        System.out.println(array[3]);
    }
}
```

运行代码，报错信息如图 3-33 所示。

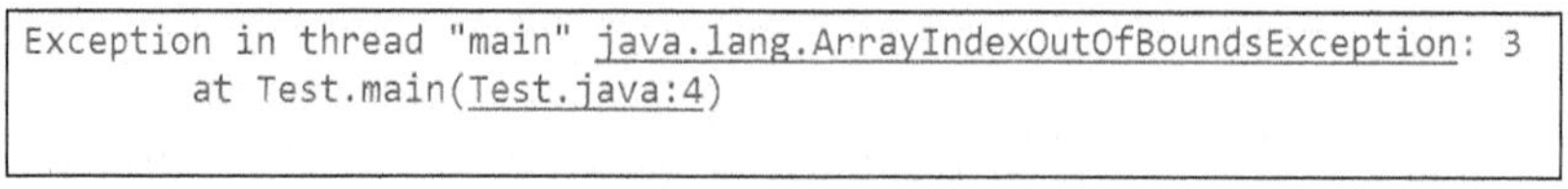

```
Exception in thread "main" java.lang.ArrayIndexOutOfBoundsException: 3
        at Test.main(Test.java:4)
```

图 3-33

java.lang.ArrayIndexOutOfBoundsException 表示错误原因是数组下标越界，array 的长度为 3，所以下标的界限为 0~2，3 已经超出这个范围。

3.3.4 数组的常用操作及方法

在实际开发中，数组的使用非常广泛，这里给大家介绍几种最常用的操作：（1）求数组中的最大值；（2）求数组中的最小值；（3）在数组的指定位置插入一个数据；（4）数组排序。求最大值和最小值的基本思路是一样的，取出数组中的第一个元素，依次与数组中的其他元素进行对比，找到目标，具体实现如代码 3-30 所示。

代码 3-30
```java
public class Test {
    public static void main(String[] args) {
        int[] array = {73,80,62,93,96,87};
        int max = array[0];
        for(int i = 1;i < array.length;i++){
            if(array[i] > max){
                max = array[i];
            }
        }
        System.out.println("最大值是"+max);
        int min = array[0];
        for(int i = 1;i < array.length;i++){
            if(array[i] < min){
                min = array[i];
            }
        }
        System.out.println("最小值是"+min);
    }
}
```

运行结果如图 3-34 所示。

图 3-34

在数组的指定位置插入一个数据：现有数组 int[] array = {96,93,87,80,73,62}，要求将 83 插入到下标为 3 的位置。

分析：初始化数组长度为 6，现要求插入一个元素。因为数组一旦创建长度是不可改变的，所以首先需要创建一个长度为 7 的新数组来存储插入之后的所有元素。然后将原数组的值复制到新数组中，同时指定位置之后的元素依次向后移动一位，最后将目标元素保存到指定位置即可，如图 3-35 所示。

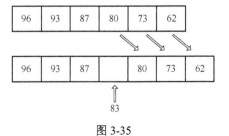

图 3-35

具体实现如代码 3-31 所示。

代码 3-31

```java
import java.util.Arrays;
public class Test {
    public static void main(String[] args) {
        //在数组下标为 3 的位置添加 83
        int[] array = {96,93,87,80,73,62};
        int score = 83;
        int[] array2 = new int[array.length+1];
        for(int i = 0; i < 3; i++){
            array2[i] = array[i];
        }
        array2[3] = 83;
        for(int i = 4;i<array2.length;i++){
            array2[i] = array[i-1];
        }
        //Arrays 工具类的 toString 方法可以将数组的元素依次取出拼接成一个字符串
        System.out.println("添加新元素之前的数组："+Arrays.toString(array));
        System.out.println("添加新元素之后的数组："+Arrays.toString(array2));
    }
}
```

运行结果如图 3-36 所示。

```
添加新元素之前的数组：[96, 93, 87, 80, 73, 62]
添加新元素之后的数组：[96, 93, 87, 83, 80, 73, 62]
```

图 3-36

数组排序是指按照升序或降序对一个数组中的所有元素进行排序，思路：依次比较数组中相邻的两个数字，如果是升序排列，则较大的数字放后面；如果是降序排列，则较大的数字放前面。这种方式也叫作冒泡排序，是一种常用算法，具体实现如代码 3-32 所示。

代码 3-32

```java
import java.util.Arrays;
public class Test {
    public static void main(String[] args) {
        int[] array = {73,80,62,93,96,87};
        //升序排列，大的放后面
        for(int j = 0; j < array.length-1;j++){
            for(int i = 0; i < array.length-1-j;i++){
                if(array[i] > array[i+1]){
                    int temp = array[i];
                    array[i] = array[i+1];
                    array[i+1] = temp;
                }
            }
        }
        System.out.println("升序排列："+Arrays.toString(array));
        //降序排列，小的放后面
        for(int j = 0; j < array.length-1;j++){
            for(int i = 0; i < array.length-1-j;i++){
                if(array[i] < array[i+1]){
```

```
            int temp = array[i];
            array[i] = array[i+1];
            array[i+1] = temp;
        }
      }
    }
    System.out.println("降序排列："+Arrays.toString(array));
  }
}
```

运行结果如图 3-37 所示。

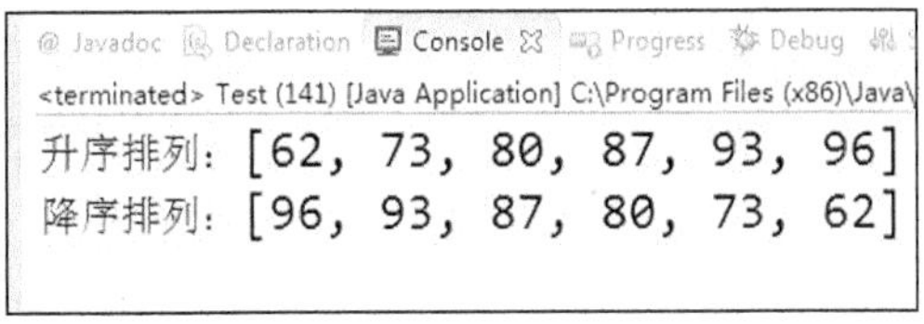

```
<terminated> Test (141) [Java Application] C:\Program Files (x86)\Java\
升序排列：[62, 73, 80, 87, 93, 96]
降序排列：[96, 93, 87, 80, 73, 62]
```

图 3-37

在实际开发中，对数组的各种操作并不需要开发者手写逻辑，Java 提供了一个工具类，通过调用该工具类的方法即可完成对数组的操作。这个类就是 Arrays，它保存在 java.util 包中，提供了多种操作数组的方法，可以让开发者更加快速、方便地进行开发。例如需要对数组进行排序，我们并不需要自己写冒泡排序，直接调用 Arrays 类的方法即可完成，Arrays 类的具体使用如代码 3-33 所示。

代码 3-33

```
import java.util.Arrays;
public class Test {
    public static void main(String[] args) {
        int[] array = {73,80,62,93,96,87};
        int[] array2 = {73,80,62,93,96,87};
        int[] array3 = {66,55,44,33,22};
        System.out.println(Arrays.equals(array, array2));
        Arrays.sort(array);
        System.out.println(Arrays.toString(array));
        Arrays.fill(array2, 66);
        System.out.println(Arrays.toString(array2));
        int[] copyArray = Arrays.copyOf(array3, 10);
        System.out.println(Arrays.toString(copyArray));
        int index = Arrays.binarySearch(array, 87);
        System.out.println(index);
    }
}
```

运行结果如图 3-38 所示。

```
true
false
[62, 73, 80, 87, 93, 96]
[66, 66, 66, 66, 66, 66]
[66, 55, 44, 33, 22, 0, 0, 0, 0, 0]
3
```

图 3-38

3.3.5 二维数组

在某些特定场景下，普通的数组已无法满足需求，如要求创建一个数组，保存 6 个货柜中所有商品的价格，每个货柜有 10 件商品。我们当然可以创建长度为 60 的数组来保存所有商品的价格，但是很显然这种方式不是特别合理，没有体现出货柜-商品的数据结构关系，不利于代码的维护和复用。这里我们可以使用二维数组来保存数据，二维数据简单理解即一维数组中保存的值也是一维数组，如图 3-39 所示。

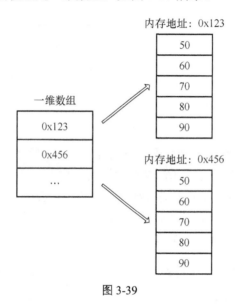

图 3-39

如果一维数组中保存的值是其他数组的内存地址，那这种结构的数组就是二维数组。二维数组的使用与一维数组类似：（1）声明；（2）开辟内存空间；（3）赋值。具体实现如代码 3-34 所示。

代码 3-34

```java
public class Test {
    public static void main(String[] args) {
        //1.声明二维数组，int[][] array 和 int array[][]都可以
        int[][] array;
        //2.开辟内存空间，第 1 个[]表示一维长度，第 2 个[]表示二维长度
        array = new int[2][3];
        //3.赋值
        array[0][0] = 1;
        array[0][1] = 2;
        array[0][2] = 3;
        array[1][0] = 4;
        array[1][1] = 5;
        array[1][2] = 6;
    }
}
```

与一维数组相同，二维数组也支持边声明边赋值的方式，具体实现如代码 3-35 所示。

代码 3-35

```
public class Test {
    public static void main(String[] args) {
        int[][] array2 = {{1,2,3},{4,5,6}};
        int[][] array3 = new int[][]{{1,2,3},{4,5,6}};
    }
}
```

例子：超市卖鸡蛋的货柜共分上、中、下 3 层，每层又分为 6 个格挡，每个格挡中的鸡蛋个数为层数×格挡数，如第 1 层的第 3 个格挡中有 1×3=3 个鸡蛋，第 3 层的第 6 个格挡种有 3×6=18 个鸡蛋，要求按照此规律使用二维数组来装载所有的鸡蛋并求出总数量，具体实现如代码 3-36 所示。

代码 3-36

```
public class Test {
    public static void main(String[] args) {
        int[][] array = new int[3][6];
        int sum = 0;
        for(int i = 0; i < array.length; i++) {
            System.out.println("---------- 第"+(i+1)+"层货架 ----------");
            for(int j = 0; j < array[i].length; j++) {
                int num = (i+1)*(j+1);
                System.out.println("第"+(j+1)+"个格挡的鸡蛋个数："+num);
                sum += num;
            }
        }
        System.out.println("鸡蛋的总数是："+sum);
    }
}
```

运行结果如图 3-40 所示。

```
---------- 第1层货架 ----------
第1个格挡的鸡蛋个数：1
第2个格挡的鸡蛋个数：2
第3个格挡的鸡蛋个数：3
第4个格挡的鸡蛋个数：4
第5个格挡的鸡蛋个数：5
第6个格挡的鸡蛋个数：6
---------- 第2层货架 ----------
第1个格挡的鸡蛋个数：2
第2个格挡的鸡蛋个数：4
第3个格挡的鸡蛋个数：6
第4个格挡的鸡蛋个数：8
第5个格挡的鸡蛋个数：10
第6个格挡的鸡蛋个数：12
---------- 第3层货架 ----------
第1个格挡的鸡蛋个数：3
第2个格挡的鸡蛋个数：6
第3个格挡的鸡蛋个数：9
第4个格挡的鸡蛋个数：12
第5个格挡的鸡蛋个数：15
第6个格挡的鸡蛋个数：18
鸡蛋的总数是：126
```

图 3-40

3.4 综合练习

使用目前所学的知识点,重点包括变量、数据类型、流程控制、循环、数组等内容,完成一个用户管理系统。需求包括:查询用户、添加用户、删除用户、账号冻结、账号解封、退出系统。

查询用户:将系统中保存的全部用户信息在控制台打印输出。

添加用户:向系统中添加新的用户信息,如果添加的用户已存在,给出提示信息。

删除用户:输入用户名称,进行删除操作,若输入的用户不存在,给出提示信息。

账号冻结:输入用户名称,进行冻结操作,若输入的用户不存在或该用户已被冻结,给出提示信息。

账号解封:输入用户名称,进行解封操作,若输入的用户不存在或该用户状态正常,给出提示信息。

退出系统:跳出循环,给出提示信息。

具体实现如代码 3-37 所示。

代码 3-37

```java
import java.util.Scanner;
public class Test {
    public static void main(String[] args) {
        //初始化用户名称
        String[] nameArray = {"张三","李四","王五","小明"};
        //初始化用户年龄
        int[] ageArray = {22,23,20,22};
        //初始化用户状态
        String[] stateArray = {"正常","正常","正常","正常"};
        Scanner scanner = new Scanner(System.in);
        int num;
        do{
            System.out.println("欢迎使用用户管理系统");
            System.out.println("1.查询用户");
            System.out.println("2.添加用户");
            System.out.println("3.删除用户");
            System.out.println("4.账号冻结");
            System.out.println("5.账号解封");
            System.out.println("6.退出系统");
            System.out.print("请选择:");
            num = scanner.nextInt();
            switch(num){
                case 1:
                    System.out.println("------查询用户------");
                    System.out.println("编号\t\t名称\t\t年龄\t\t状态");
                    for(int i = 0; i < nameArray.length;i++){
                        if(nameArray[i] != null){

                            System.out.println((i+1)+"\t\t"+nameArray[i]+"\t\t"+ageAr
```

```java
ray[i]+"\t\t"+stateArray[i]);
                    }
                }
                System.out.print("输入 0 返回：");
                num = scanner.nextInt();
                break;
            case 2:
                System.out.println("------添加用户------");
                //判断数组是否已满
                if(nameArray[nameArray.length-1] != null){
                    //更新数组长度
                    String[] newNameArray = new String[nameArray.length+1];
                    String[] newStateArray = new String[stateArray.length+1];
                    int[] newAgeArray = new int[ageArray.length+1];
                    for(int i = 0;i < nameArray.length;i++) {
                        newNameArray[i] = nameArray[i];
                        newStateArray[i] = stateArray[i];
                        newAgeArray[i] = ageArray[i];
                    }
                    nameArray = newNameArray;
                    stateArray = newStateArray;
                    ageArray = newAgeArray;
                }
                System.out.print("请输入用户名称：");
                String name = scanner.next();
                boolean flag = false;
                //判断该用户是否存在
                for(int i = 0; i < nameArray.length; i++){
                    if(nameArray[i] != null && nameArray[i].equals(name)){
                        System.out.println(name+"已存在！");
                        flag = true;
                        break;
                    }
                }
                //添加用户
                if(!flag){
                    System.out.print("请输入用户年龄：");
                    int age = scanner.nextInt();
                    nameArray[nameArray.length-1] = name;
                    stateArray[stateArray.length-1] = "正常";
                    ageArray[ageArray.length-1] = age;
                    System.out.println(name+"添加成功！");
                }
                System.out.print("输入 0 返回：");
                num = scanner.nextInt();
                break;
            case 3:
                System.out.println("------删除用户------");
                System.out.print("请输入用户名称：");
                name = scanner.next();
                //判断该用户是否存在
                boolean flag2 = false;
                for(int i = 0; i < nameArray.length; i++){
                    if(nameArray[i] != null && nameArray[i].equals(name)){
                        flag2 = true;
                        if(i == nameArray.length-1 ){
                            nameArray[i] = null;
                            stateArray[i] = null;
```

```java
                }else{
                    for(int j = i;j<nameArray.length-1;j++){
                        nameArray[j] = nameArray[j+1];
                        stateArray[j] = stateArray[j+1];
                        nameArray[j+1] = null;
                        stateArray[j+1] = null;
                    }
                }
            }
        }
        if(!flag2){
            System.out.println(name+"不存在,请重新输入!");
        }else{
            System.out.println(name+"删除成功!");
        }
        System.out.print("输入 0 返回: ");
        num = scanner.nextInt();
        break;
    case 4:
        System.out.println("------账号冻结------");
        System.out.print("请输入用户名称: ");
        name = scanner.next();
        //判断该用户是否存在
        boolean flag3 = false;
        for(int i = 0; i < nameArray.length; i++){
            if(nameArray[i] != null && nameArray[i].equals(name)){
                flag3 = true;
                if(stateArray[i].equals("冻结")){
                    System.out.println(name+"已冻结!");
                }else{
                    stateArray[i] = "冻结";
                    System.out.println(name+"冻结成功!");
                }
                break;
            }
        }
        if(!flag3){
            System.out.println(name+"不存在,请重新输入!");
        }
        System.out.print("输入 0 返回: ");
        num = scanner.nextInt();
        break;
    case 5:
        System.out.println("------账号解封------");
        System.out.print("请输入用户名称: ");
        name = scanner.next();
        //判断该用户是否存在
        boolean flag4 = false;
        for(int i = 0; i < nameArray.length; i++){
            if(nameArray[i] != null && nameArray[i].equals(name)){
                flag4 = true;
                if(stateArray[i].equals("正常")){
                    System.out.println(name+"状态正常!");
                }else{
                    stateArray[i] = "正常";
                    System.out.println(name+"解封成功!");
                }
                break;
```

```
            }
        }
        if(!flag4){
            System.out.println(name+"不存在，请重新输入！");
        }
        System.out.print("输入 0 返回：");
        num = scanner.nextInt();
        break;
    case 6:
        System.out.println("感谢使用用户管理系统！");
        return;
    }
}while(num == 0);
    }
}
```

运行结果如图 3-41～图 3-52 所示。

```
欢迎使用用户管理系统
1.查询用户
2.添加用户
3.删除用户
4.账号冻结
5.账号解封
6.退出系统
请选择：1
------查询用户------
编号        名称        年龄        状态
1          张三        22         正常
2          李四        23         正常
3          王五        20         正常
4          小明        22         正常
输入0返回：
```

图 3-41

```
欢迎使用用户管理系统
1.查询用户
2.添加用户
3.删除用户
4.账号冻结
5.账号解封
6.退出系统
请选择：2
------添加用户------
请输入用户名称：张三
张三已存在！
输入0返回：
```

图 3-42

```
欢迎使用用户管理系统
1.查询用户
2.添加用户
3.删除用户
4.账号冻结
5.账号解封
6.退出系统
请选择：2
------添加用户------
请输入用户名称：小红
请输入用户年龄：22
小红添加成功！
输入0返回：
```

图 3-43

```
欢迎使用用户管理系统
1.查询用户
2.添加用户
3.删除用户
4.账号冻结
5.账号解封
6.退出系统
请选择：1
------查询用户------
编号        名称        年龄        状态
1          张三         22         正常
2          李四         23         正常
3          王五         20         正常
4          小明         22         正常
5          小红         22         正常
输入0返回：
```

图 3-44

```
欢迎使用用户管理系统
1.查询用户
2.添加用户
3.删除用户
4.账号冻结
5.账号解封
6.退出系统
请选择：4
------账号冻结------
请输入用户名称：小黑
小黑不存在，请重新输入！
输入0返回：
```

图 3-45

```
欢迎使用用户管理系统
1.查询用户
2.添加用户
3.删除用户
4.账号冻结
5.账号解封
6.退出系统
请选择：4
------账号冻结------
请输入用户名称：小红
小红冻结成功！
输入0返回：
```

图 3-46

```
欢迎使用用户管理系统
1.查询用户
2.添加用户
3.删除用户
4.账号冻结
5.账号解封
6.退出系统
请选择：1
------查询用户------
编号        名称        年龄        状态
1          张三         22         正常
2          李四         23         正常
3          王五         20         正常
4          小明         22         正常
5          小红         22         冻结
输入0返回：
```

图 3-47

```
欢迎使用用户管理系统
1.查询用户
2.添加用户
3.删除用户
4.账号冻结
5.账号解封
6.退出系统
请选择：4
------账号冻结------
请输入用户名称：小红
小红已冻结！
输入0返回：
```

图 3-48

```
欢迎使用用户管理系统
1.查询用户
2.添加用户
3.删除用户
4.账号冻结
5.账号解封
6.退出系统
请选择：5
------账号解封------
请输入用户名称：张三
张三状态正常！
输入0返回：
```

图 3-49

```
欢迎使用用户管理系统
1.查询用户
2.添加用户
3.删除用户
4.账号冻结
5.账号解封
6.退出系统
请选择：5
------账号解封------
请输入用户名称：小红
小红解封成功！
输入0返回：
```

图 3-50

```
欢迎使用用户管理系统
1.查询用户
2.添加用户
3.删除用户
4.账号冻结
5.账号解封
6.退出系统
请选择：3
------删除用户------
请输入用户名称：小红
小红删除成功！
输入0返回：
```

图 3-51

```
欢迎使用用户管理系统
1.查询用户
2.添加用户
3.删除用户
4.账号冻结
5.账号解封
6.退出系统
请选择：6
感谢使用用户管理系统！
```

图 3-52

3.5 小结

本章为大家讲解了 Java 中的流程控制和循环等基本逻辑处理，使用这些技术编写的 Java 程序不再只是简单的数据展示，而是具备了一定的任务处理能力。本章还为大家讲解了数组的使用，在实际开发中数据量比较大的情况下，使用数组来管理数据可以简化代码。

第2部分　Java 面向对象

第 4 章 面向对象基础

> 在 Java 中有句俗语叫作"万物皆对象",这句话的意思是我们可以把 Java 程序看成是很多个对象构建的系统。具体来说就是将 Java 程序的所有参与角色都看成一个个对象,通过对象与对象之间的相互调用来完成系统的功能。这是一种将程序模块化的思想,叫作面向对象编程思想。不仅仅是 Java,C++、C#等高级编程语言也属于面向对象的范畴。那么到底什么是面向对象?除了面向对象还有哪些编程思想?为什么要采用面向对象这种编程思想呢?我们来一探究竟。

4.1 什么是面向对象

在面向对面编程思想问世之前,程序开发采用的是面向过程的结构化编程方式,这是一种面向功能划分的软件结构。自上向下,将一个大问题分解成几个小问题,再将小问题分解成更小的问题,最后将任务划分成一个一个步骤,然后按照步骤分别去执行,最小粒度细化到了方法这一层面。这种方式对于项目开发会有很多局限性,因为是逐步执行的,所以开发步骤非常烦琐,同时制约了程序的可维护性和可扩展性。这么说很抽象,我们通过一个现实生活中的例子来解释它:假如你现在需要开车去北京,如果按照面向过程的思想来完成这一需求,就需要将开车去北京这件事情细化成一个个具体的行为,并且按顺序去执行。关注点在于每一步操作,那么我们就需要这样处理:第 1 步,打开车门;第 2 步,坐进驾驶室;第 3 步,发动汽车;第 4 步,踩油门出发……第 n 步,遇到红灯停车……整个过程的每一步都需要记录下来,非常麻烦,相当于你需要亲力亲为把车开到北京。

如果用面向对象的思路来处理这个需求会有怎么样的效果呢?没有对比就没有伤害,你会发现使用面向对象的思想来处理,这件事情就变得非常简单。面向对象就是将程序的参与者全部模块化成对象,这件事情可以模块化出来 3 个对象:你、汽车、北京。然后只需要让这 3 个对象相互调用就可以完成需求:让汽车载着你到达北京,忽略了过程中每一个细节,关注点在 3 个对象之间的关系。这是与面向过程完全不同的一种思维方式,面向过程注重的是每一个步骤,面向对象关注点在于整件事情的模块化结构。

我们为什么要采用面向对象这种编程思想呢?其实上面已经讲到了,面向对象的核心思想就是重用性以及灵活性（灵活性=可扩展性+变化性）。重用性是指已经写好的某个业务代码，可以在多个不同的功能模块中复用，而不必针对每个功能模块都去编写相同的代码。灵活性是指当某个功能模块完成之后，根据需求的变化可以迅速在原有代码基础上更新业务逻辑，以实现新的需求。

再回到 Java，我们说在 Java 中万物皆对象，其实是在说 Java 是一门面向对象的编程语言，Java 程序的功能都是通过对象来完成的，根据业务需求将程序分解为不同的对象，再通过对象之间的相互调用关系来协同实现相应的功能。对编程思想的理解会因为你做的业务和代码的累积发生改变，随着知识的深入以及视野的开阔，一段时间后再回头看目前你所掌握的这些东西，你可能会发现某些地方理解得不够透彻甚至有误，这就是学习的过程。

4.2 类与对象

4.2.1 类与对象的关系

在 Java 的世界中，我们可以把世间万物都看作对象，例如一个人、一间房子、一台电脑等，总之任何事物我们都可以将其看作对象，然后通过对象之间的相互调用来完成需求。既然对象可以描述任何事物，那么不同的对象一定是有差别的，每一个对象都应该有自己的特征，如何来描述这些特征呢？（1）属性。（2）方法。属性指的是对象的静态特征，例如一个人的年龄、身高、性别等，一辆车的颜色、品牌、排量等。方法用来描述对象的动态特征，例如一个人可以跑步、说话、吃饭，一辆车可以加速、刹车，一静一动两方面特征就构成了一个完整独立的对象。总结一下：对象就是用来描述客观存在的一个实体，该实体由一组属性和方法构成。

与对象紧密结合的另外一个概念是类，关于类，我们之前有过简单的介绍。我们说过类是用来组织 Java 程序的，一个 Java 文件就是一个类，这是从物理属性的层面来解读的，也是最基本的认识，这里需要对类有更深层次的理解，一句话来解释：类是产生对象的模板，所有的对象都是通过类来创建的。二者关系：类是对象的抽象化描述，这些对象具有相同的特征和动作（属性和方法），而对象是类的具体实例。例如狗是类，那么隔壁家的小狗"旺财"就是对象；例如计算机是类，那么你面前的这台计算机就是对象；例如学生是类，那么教室中的每一位同学就是对象，如图 4-1 所示。

类可以创建出很多个对象，一般来讲一个对象归属于一个类，但是不够准确，例如我可以说旺财是狗，也可以说旺财是宠物，可以说我现在用的笔记本电脑是一台电脑，也可以说它是一个电子产品。类是抽象概念，是一种描述，仅仅是模板，对象是实实在在的具体存在的，对象和类是面向对象编程思想的核心。所有的 Java 程序都是以类为组织单元，程序运行时的主体是通过类创建的具体化对象。例如学生来上课，我们需要定义学生类、班级类和课程类。这 3 个类定义好之后，每一次学生来上课的业务场景都可以用这 3 个类产生 3 个具体对象来完成。

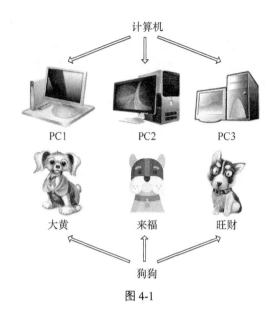

图 4-1

（1）张三去 1 班上 Java 课。

（2）李四去 2 班上大数据课。

……

这个例子中对象是可以产生很多个的，没有上限，但是类只有 3 个。

4.2.2 定义类

了解了类的概念以及类与对象的关系，接下来我们学习如何定义一个类，基本语法：

```
public class 类名{
//定义属性，属性名符合驼峰式命名法
public 数据类型 属性名；
//定义方法，方法名符合驼峰式命名法
public 返回值类型 方法名(参数列表){
//方法体
}
}
```

定义属性需要指定其数据类型，语法比较简单，方法的定义比属性稍复杂一些，有两个重点：返回值类型和参数列表。定义方法需要指定方法的返回值类型，Java 关于返回值的定义分为两类：有返回值和无返回值。有返回值的方法需要在方法定义时指定返回值数据类型，并在方法体中用 return 将结果返回给外部调用者，如一个加法运算的方法，将 10+10 的结果返回给外部调用者，如代码 4-1 所示。

代码 4-1

```
public int add(){
```

```
        return 10+10;
    }
```

参数列表是指外部在调用该方法时需要传到方法内部进行运算的数据,在方法名后面的括号中定义。语法:数据类型 参数名,我们对上述代码进行优化,将外部调用 add 方法时传入的两个参数进行相加,然后返回结果,具体实现如代码 4-2 所示。

代码 4-2

```java
public int add(int num1,int num2){
    return num1+num2;
}
```

如果一个方法不需要进行返回操作,我们将其返回值类型定义为 void,这里需要强调没有返回值的方法就用 void 来修饰其返回值类型,如我们定义一个方法打印"Hello World",不需要返回值,如代码 4-3 所示。

代码 4-3

```java
public void add(){
    System.out.println("Hello World");
}
```

类名一般要首字母大写,驼峰式命名法是指第一个单词全部小写,后续的单词仅首字母大写,如 userName、myScore。定义一个用户类,具体实现如代码 4-4 所示。

代码 4-4

```java
public class User {
    //定义属性
    public int id;
    public String name;
    public char gender;
    public String password;
    //定义方法
    public void show(){
        System.out.println("展示用户信息");
    }
}
```

4.2.3 构造函数

现在我们已经学会了如何定义一个类,接下来就可以通过类来创建对象了,如何完成呢?Java 是通过类的构造函数,也叫构造方法来创建对象的。构造函数是一种特殊的方法,方法名必须与类名一致,不需要定义返回值类型,基本语法如下。

```
public 构造函数名(参数列表){
}
```

构造函数包括有参构造和无参构造,我们来创建 User 类的构造函数,如代码 4-5

代码 4-5

```java
public class User {
    //定义属性
    public int id;
    public String name;
    public char gender;
    public String password;
    //无参构造函数
    public User(){}
    //有参构造函数
    public User(int id,String name,char gender,String password ){
        this.id = id;
        this.name = name;
        this.gender = gender;
        this.password = password;
    }
}
```

每个类默认都有一个无参构造函数，我们在定义类时不需要声明无参构造函数，便可直接调用来创建对象，但是如果手动定义一个有参构造函数，就会覆盖掉默认的无参构造函数。

4.2.4 创建对象

创建对象只需要调用对应类的构造函数即可，构造函数分为无参构造和有参构造，区别在于调用无参构造创建的对象，不会给属性赋值，需要手动对属性进行赋值，先创建再赋值，如代码 4-6 所示。

代码 4-6

```java
public class Test {
    public static void main(String[] args) {
        //创建一个 User 对象, id:1, name:张三, gender:男, password:root
        User user = new User();
        user.id = 1;
        user.name = "张三";
        user.gender = '男';
        user.password = "root";
    }
}
```

调用有参构造，则不需要在构造函数外部进行属性赋值，构造函数本身就会为创建好的对象赋值，只需要在调用有参构造时将属性值作为参数传入即可，边创建边赋值，如代码 4-7 所示。

代码 4-7

```java
public class Test {
    public static void main(String[] args) {
```

```
        //边创建边赋值
        User user = new User(1,"张三",'男',"root");
    }
}
```

4.2.5 使用对象

对象的使用包括获取和修改属性，以及调用方法。访问属性通过"对象名.属性名"来完成，调用方法通过"对象名.方法名(参数列表)"来完成，如代码 4-8 所示。

代码 4-8

```
public class Test {
    public static void main(String[] args) {
        User user = new User(1,"张三",'男',"root");
        System.out.println("用户姓名："+user.name);
        user.id = 2;
        System.out.println("用户编号："+user.id);
        user.show();
    }
}
```

运行结果如图 4-2 所示。

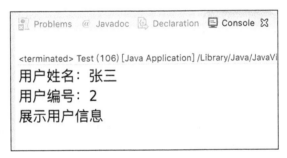

图 4-2

4.2.6 this 关键字

在类的定义中我们通常会使用 this 关键字，this 用来指代当前类的实例化对象，通过 this 可以调用当前类的属性和方法，比如在有参构造函数中，通过 this 将外部传来的值赋给当前类的实例化对象，如图 4-3 所示。

```
            //有参构造函数
            public User (int id,String name,char gender,String password){
                this.id = id;
当前类的属性     this.name = name;
                this.gender = gender;
                this.password = password;    将外部传入的值赋给当前类的属性
            }
```

图 4-3

this 除了可以在类中访问属性也可以在类中调用方法，我们知道类中的方法可以分为两类：构造方法和普通方法。用 this 调用这两类方法的语法也不相同，调用构造函数的语法是"this(参数列表);"，不能在普通方法中使用 this 来调用构造函数，调用普通方法的语法是"this.方法名(参数列表)"，在构造方法中可以使用 this 来调用普通方法，如代码 4-9 所示。

代码 4-9

```java
public class User {
    public User(int id,String name,char gender,String password ){
        //调用无参构造函数
        this();
        //调用普通方法
        this.show();
    }
}
```

4.2.7 方法重载

方法重载是 Java 代码复用的一种重要方式，指的是两个方法之间的一种关系，那么方法之间具备什么条件就可以构成重载呢？

（1）同一个类中。

（2）方法名相同。

（3）参数列表不同（个数或类型不同）。

（4）与返回值和访问权限修饰符无关。

接下来我们看一个具体的应用，先定义一个 Test 类，然后定义两个 method 方法，一个有参数，一个没有参数，具体实现如代码 4-10 所示。

代码 4-10

```java
public class Test {
    public static void main(String[] args) {
        Test test = new Test();
        test.method();
        test.method(10);
    }
    public void method() {
        System.out.println("没有参数");
    }
    public void method(int num) {
        System.out.println("参数为："+num);
    }
}
```

运行结果如图 4-4 所示。

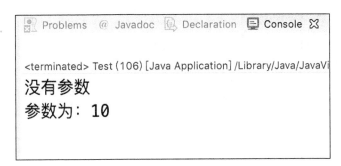

图 4-4

例子,以下哪种情况属于方法重载?

```
A.public int test(){}
  public int test(){}

B.public double test(double a,double b){}
  public double test(){}

C.public String test(){}
  public void test(int a){}

D.public void test(){}
  public void test2(){}
```

B 和 C 属于方法重载,A 中两个方法的参数列表相同,D 中两个方法名不同,所以均不是方法重载。

4.2.8 成员变量和局部变量

我们知道变量是 Java 程序中表示数据的基本单位,两个变量之间除了数据类型、变量名和变量值这些内容不同之外,变量的作用域范围也是不同的。变量的作用域是指在程序中可以通过变量名来访问该变量的范围,变量的作用域由变量被声明时所在的位置决定,Java 中根据不同的作用域可以将变量分为成员变量和局部变量。如何判断一个变量是成员变量还是局部变量?很简单,根据该变量的声明位置即可得出结论,如果一个变量在方法中声明,则该变量为局部变量。如果一个变量在方法外,在类中声明,则该变量为成员变量,如图 4-5 所示。

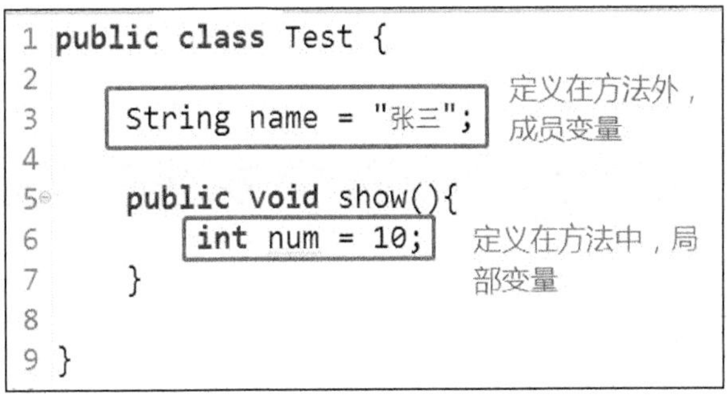

图 4-5

成员变量和局部变量的作用域有什么区别呢？成员变量的作用域在整个类中，类中的每个方法中都可以访问该变量。局部变量的作用域只在定义该变量的方法中，出了方法体就无法访问，如图 4-6 所示。

```
 1  public class Test {
 2      String name = "张三";
 3      public void show(){
 4          int num = 10;
 5          System.out.println(name);
 6          System.out.println(num);
 7      }
 8      public void show2(){
 9          System.out.println(name);
10          System.out.println(num);
11      }
12  }
```
num cannot be resolved to a variable
4 quick fixes available:

图 4-6

在 show 方法中，name 和 num 都可以访问，而在 show2 方法中，name 可以访问而 num 不能访问。因为 name 是成员变量，Test 类的每个方法都可以访问该变量。num 在 show 方法中定义，是局部变量，作用域仅局限在 show 方法中，一旦出了 show 方法就无法访问。

例子，下列代码的运行结果是什么？

```java
public class Test {
    int num = 9;
    public void show(){
        int num = 10;
        System.out.println(num);
    }
    public static void main(String[] args) {
        Test test = new Test();
        test.show();
    }
}
```

运行结果为 10。当成员变量和局部变量重名时局部变量的优先级更高。这里特别说明一下，Java 程序运行的入口是 main 方法，每个类都可以定义 main 方法，即每个类都可以作为 Java 程序运行的入口，要调用类中的方法，就必须首先实例化类的对象，然后通过对象来调用方法，例如 "test.show();"。现在我们对代码进行修改，如代码 4-11 所示。

代码 4-11

```java
public class Test {
    int num = 9;
    public void show(){
        int num = 10;
        System.out.println(this.num);
    }
```

```
    public static void main(String[] args) {
        Test test = new Test();
        test.show();
    }
}
```

此时，调用 show 方法的运行结果是 9，this 指当前类的实例化对象，所以 this.num 访问的是当前类的成员变量 num，作用域是整个类，而非局部变量 num。成员变量和局部变量除了作用域不同之外初始值也不同，Java 会给成员变量赋初始值，局部变量则不会赋初始值，不同类型的成员变量的初始值不同，如代码 4-12 所示。

代码 4-12

```
public class Test {
    String name;
    byte num1;
    int num2;
    short num3;
    long num4;
    double num5;
    float num6;
    char num7;
    boolean num8;
    public void show(){
        System.out.println(name);
        //依次打印其他属性
    }
    public static void main(String[] args) {
        Test test = new Test();
        test.show();
    }
}
```

运行结果如图 4-7 所示。

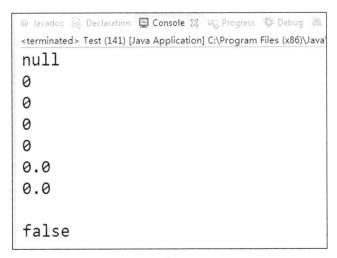

图 4-7

4.3 封装

4.3.1 什么是封装

在学习封装的概念之前，我们先来了解一下封装的背景，即为什么要有封装。我们知道类中可以定义属性，用来描述类的静态特征。通过类创建对应的实例化对象，可以对属性进行访问和修改，我们来看下面这个例子，这里定义了一个 User 类，如代码 4-13 所示。

代码 4-13

```java
public class User {
    public int id;
    public String name;
    public char gender;
    public String password;
    public void show(){
        System.out.println("用户信息如下：");
        System.out.println("用户编号："+this.id);
        System.out.println("用户姓名："+this.name);
        System.out.println("用户性别："+this.gender);
        System.out.println("用户密码："+this.password);
    }
}
```

测试类的 main 方法作为程序的入口，实例化 User 对象并且给属性赋值，同时打印对象信息，如代码 4-14 所示。

代码 4-14

```java
public class Test {
    public static void main(String[] args) {
        User user = new User();
        user.id = -1;
        user.name = "张三";
        user.gender = '男';
        user.password = "root";
        user.show();
    }
}
```

程序运行结果如图 4-8 所示。

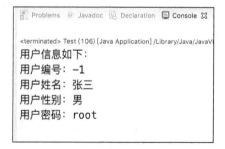

图 4-8

从程序的角度来讲，这段代码是没有问题的，但是从现实逻辑来讲是有 bug 的。因为用户编号不可能为负数，所以我们之前定义类的方式就存在一个漏洞，即外部在实例化类对象时，可以随意给对象的属性赋值，只要数据类型一致即可。这样就会存在现实逻辑可能出 bug 的隐患。那么问题来了，这个隐患怎么解决呢？

我们可以在类中对属性的赋值加以限制，将外部传来的值进行筛选，合格的完成赋值，不合格的加以处理，这个过程就是封装。封装是指将类的属性隐藏在内部，外部不能直接访问和修改，必须通过类提供的方法来完成对属性的访问和修改。封装的核心思想就是尽可能把属性都隐藏在内部，对外提供方法来访问，我们可以在这些方法中添加逻辑处理来实现过滤，以屏蔽错误数据的赋值。

4.3.2 封装的步骤

了解完封装的概念，接下来我们学习如何使用封装，分 3 步操作：(1) 修改属性的访问权限，使得外部不能够直接访问；(2) 提供外部可以直接调用的方法；(3) 在方法中加入属性控制逻辑。

修改属性的访问权限，那什么是访问权限？访问权限指的是该属性可以被直接访问的范围，是在属性定义时设定的。在我们之前所写的代码中，属性的访问权限为公有（public），即该属性是公开的，在外部可以直接访问，如图 4-9 所示。

```
public class User {
    public int id;
    public String name;
    public char gender;
    public String password;

    public void show() {
        System.out.println("用户信息如下: ");
        System.out.println("用户编号: "+this.id);
        System.out.println("用户姓名: "+this.name);
        System.out.println("用户性别: "+this.gender);
        System.out.println("用户密码: "+this.password);
    }
}
```

图 4-9

现在我们需要将访问权限设置为私有（private），即只能在类的内部访问，外部无法直接访问该属性，将 User 类中所有属性的访问权限改为 private，如代码 4-15 所示。

代码 4-15

```
public class User {
    private int id;
    private String name;
    private char gender;
    private String password;
    ……
}
```

此时，Test 类中的代码会立即报错，如图 4-10 所示。

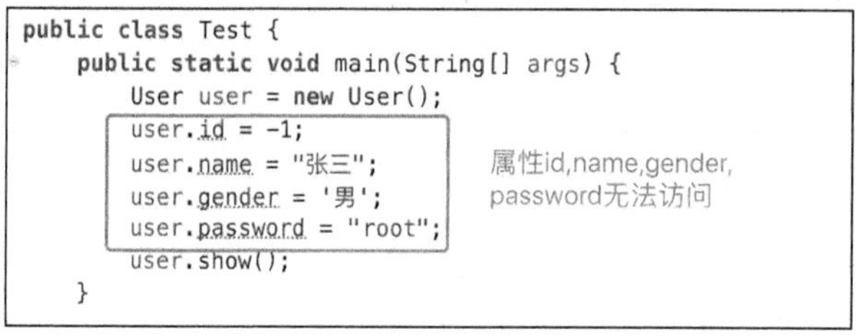

```
public class Test {
    public static void main(String[] args) {
        User user = new User();
        user.id = -1;
        user.name = "张三";          属性id,name,gender,
        user.gender = '男';           password无法访问
        user.password = "root";
        user.show();
    }
}
```

图 4-10

第 1 步我们已经完成，现在外部不能直接访问 User 的属性。接下来完成第 2 步，提供外部可以调用的方法，方法同样需要设置访问权限，很显然需要将方法的访问权限设置为 public。同时完成第 3 步，在方法中添加确保属性值正确的逻辑代码，修改完成如代码 4-16 所示。

代码 4-16

```java
public class User {
    ......
    public int getId() {
        return id;
    }
    public void setId(int id) {
        //如果外部传入的 id 小于 0，默认赋值为 1
        if(id < 0){
            this.id = 1;
        }else {
            this.id = id;
        }
    }
    //其他属性的 getter、setter 方法
}
```

封装完成，每一个属性都添加了两个方法：getter 和 setter，外部通过 getter 获取该属性的值，通过 setter 方法修改该属性的值，Test 的代码会做相应的修改，如代码 4-17 所示。

代码 4-17

```java
public class Test {
    public static void main(String[] args) {
        User user = new User();
        user.setId(-1);
        user.setName("张三");
        user.setGender('男');
        user.setPassword("root");
        user.show();
    }
}
```

运行结果如图 4-11 所示。

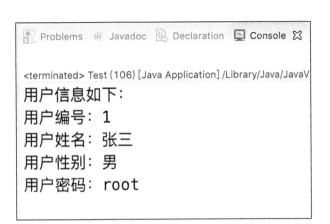

图 4-11

4.3.3 static 关键字

Java 程序要运行就必须在某个类中定义一个 main 方法作为程序的入口，我们会发现在定义 main 方法时，有一个 static 关键字，那 static 表示什么呢？为什么要用 static 来修饰 main 方法？本节内容为你一一揭晓。

static 表示静态或全局，可以用来修饰成员变量和成员方法以及代码块。要访问类中的成员变量或者成员方法，必须首先创建该类的对象，然后通过对象才能访问成员变量和成员方法，也就是说成员变量和成员方法的访问必须依赖于对象，这是常规的方式。而使用 static 修饰的成员变量和成员方法独立于该类的任何一个实例对象，访问时不依赖于该类的对象，可以理解为被该类的所有实例对象共用，所以说它是全局。用 static 修饰的成员变量叫作静态变量也叫作类变量，static 修饰的成员方法叫作静态方法，也叫作类方法。多个对象共用，内存中只有一份，没有被 static 修饰的成员变量叫实例变量，没有被 static 修饰的成员方法叫作实例方法，一个对象对应一个，内存中有多份，如图 4-12 所示。

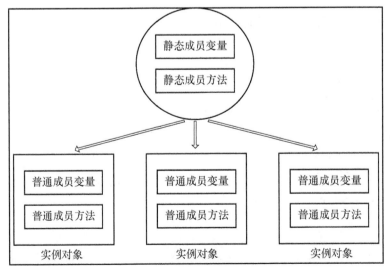

图 4-12

既然静态成员变量和静态成员方法独立于任何一个实例对象,我们在访问这些静态资源时就无须创建对象,直接通过类即可访问,如代码 4-18 所示。

代码 4-18
```
public class User {
    public static String name;
    public static void show(){
        System.out.println("这是一个 User 对象");
    }
}
```

测试类的 main 方法作为程序的入口,直接通过类访问静态成员变量和静态成员方法如代码 4-19 所示。

代码 4-19
```
public class Test {
    public static void main(String[] args) {
        User.name = "张三";
        System.out.println(User.name);
        User.show();
    }
}
```

程序运行结果如图 4-13 所示。

```
@ Javadoc  Declaration  Console ⊠  Progress  Debug
<terminated> Test (141) [Java Application] C:\Program Files (x86)\Java\
张三
这是一个User对象
```

图 4-13

例子,请给出下列 3 段代码的运行结果。

代码 A:

```
public class Test {
    private int id;
    public static void main(String[] args) {
        Test test = null;
        for(int i = 0; i < 10; i++) {
            test = new Test();
            test.id++;
        }
        System.out.println(test.id);
    }
}
```

代码 A 的运行结果为 1,因为此时的 id 为成员变量,每个实例化对象都拥有一个自己的 id。循环中创建了 10 个 Test 对象,内存中就会有 10 个 id,每个 id 都进行了加一操作，由 0 变为 1，并且每次循环都会将新创建的 Test 对象赋给变量 test，相当于更新了 10 次 test 的引用。因为每次的 id 值都为 1，所以最终的结果也为 1。

代码 B：

```java
public class Test {
    private static int id;
    public static void main(String[] args) {
        Test test = null;
        for(int i = 0; i < 10; i++) {
            test = new Test();
            test.id++;
        }
        System.out.println(test.id);
    }
}
```

代码 B 的运行结果为 10，因为此时的 id 是静态变量，属于类不属于对象，内存中只有一份，所以任何时候对 id 的修改都会作用于同一个变量，循环中共执行了 10 次 num++，所以 num 的结果为 10。

代码 C：

```java
public class Test {
    private static int id;
    public static void main(String[] args) {
        for(int i = 0; i < 10; i++) {
            Test.id++;
        }
        System.out.println(Test.id);
    }
}
```

代码 C 的运行结果为 10，具体的原因同代码 B，唯一的区别在于代码 B 是通过实例化对象来访问静态变量的，代码 C 是通过类来访问静态变量的，运行结果没有区别。

我们在使用 static 修饰成员方法时需要注意，静态方法中不能使用 this 关键字，不能直接访问所属类的实例变量和实例方法，可直接访问类的静态变量和静态方法。若要访问类的实例变量和实例方法，必须先实例化类的对象，然后通过对象去访问，如代码 4-20 所示。

代码 4-20

```java
public class User {
    public static String name;
    public int num;
    public void test(){}
    public static void show(){
        System.out.println("这是一个 User 对象");
        User user = new User();
        user.num=1;
        user.test();
    }
}
```

}

　　static 除了可以修饰成员变量和成员方法之外，还可以修饰代码块，被 static 修饰的代码块叫作静态代码块。静态代码块的特点是只执行一次，在什么时候执行呢？当该类被加载到内存时执行，不需要手动调用，它会自动执行，在什么时候类第一次被加载？这里我们需要简单地阐述一下 Java 加载类的机制。我们知道 Java 代码是由类构成的，但是真正运行时是通过对象和对象之间的相互调用关系来完成需求的。即程序运行时，需要在内存中创建多个对象，对象怎么创建？需要通过类来创建，类是模板，同时这个模板只需要加载一次，所以程序在运行时，首先会将程序中用到的类加载到内存中，并且只加载一次。然后通过类来创建多个对象以完成具体的业务。被加载到内存中的类叫作运行时类，静态代码块就是在加载类的时候执行的，因为类只加载一次，所以静态代码块也只执行一次。简单理解，当代码中第一次出现某个类时，就会执行静态代码块，静态代码块只能访问静态变量和静态方法，静态代码块的定义如代码 4-21 所示。

代码 4-21

```
public class User {
    public static int num;
    static{
        num++;
        System.out.println("执行了静态代码块");
    }
}
```

测试类的 main 方法作为程序的入口，如代码 4-22 所示。

代码 4-22

```
public class Test {
    public static void main(String[] args) {
        User user = new User();
        User user2 = new User();
        User user3 = new User();
        System.out.println(user.num);
        System.out.println(user2.num);
        System.out.println(user3.num);
    }
}
```

程序运行结果如图 4-14 所示。

```
执行了静态代码块
1
1
1
```

图 4-14

通过结果可以得出结论：虽然创建了 3 个 User 对象，但是静态代码块只执行了一次。如果有多个静态块同时存在，则按先后顺序执行，类的构造方法用于初始化类的实例，类的静态代码块用于初始化类，给类的静态变量赋值。

4.4 继承

4.4.1 什么是继承

在讲解继承的概念之前，我们先来看一个示例，定义一个 Student 类和一个 Teacher 类，分别有 id、name、age、gender 属性，如代码 4-23 所示。

代码 4-23
```java
public class Student {
    private int id;
    private String name;
    private int age;
    private char gender;
    //getter、setter 方法
}

public class Teacher {
    private int id;
    private String name;
    private int age;
    private char gender;
    //getter、setter 方法
}
```

可以看到两个类中的属性完全一样，提供给外部调用的 setter 和 getter 方法也完全一致。我们要养成一个思维习惯，当看到代码中有完全重复的内容时就需要想办法进行优化，能不能将两个类中完全一致的内容提取出来，同时让这两个类来复用这些代码呢？来找找 Student 和 Teacher 的共性，我们可不可以定义一个 People 类，然后让 Student 和 Teacher 拥有 People 类的属性和方法呢？这种代码优化的方式叫作继承，即一个类继承另外一个类的属性和方法，被继承的类叫父类，继承的类叫子类。People 就是父类，Student 和 Teacher 就是子类。那继承如何实现呢？继承的基本语法如下：

父类：
```
public class 类名{
//属性和方法
}
```

子类：
```
public class 类名 extends 父类名{
```

```
    //子类特有的属性和方法
}
```

如代码 4-24 所示。

代码 4-24

```java
public class People {
    private int id;
    private String name;
    private int age;
    private char gender;
    //getter、setter 方法
}


public class Student extends People {
}


public class Teacher extends People {
}
```

继承的好处是我们只需要定义一个父类 People，然后让 Student 和 Teacher 直接继承 People，Student 和 Teacher 中就不需要定义属性和方法了，而直接拥有了 People 的公有属性和方法。若子类中有特定的属性和方法，则只需要在继承的基础上，在子类中定义特有的属性和方法即可，此时子类的信息由两部分内容组成，一部分是继承自父类的属性和方法，另外一部分是自己特有的属性和方法。和现实生活中的例子是一样的，儿子可以继承父亲的资产，那么儿子就不需要那么辛苦打拼，可以轻松拥有父亲给他的一切，同时儿子还可以在继承父亲资产的基础上继续创造属于自己的资产。

继承是面向对象编程思想的主要特征，Java 通过继承可以实现代码复用。Java 只支持单继承，即一个类只能有一个直接父类。注意，我们这里说的是只能有一个直接父类，父类的父类资源也是可以被继承的，相当于父亲从爷爷那里继承的资产，可以传到儿子手上。

4.4.2 子类访问父类

实现了继承关系的父子类，在创建子类对象时，无论调用无参构造还是有参构造，都会默认先创建父类对象，并且是通过父类的无参构造完成实例化的，如代码 4-25 所示，在父类 People 和子类 Student 的构造函数中分别打印相关信息，在测试类的 main 方法中分别调用 Student 的无参构造和有参构造来创建 Student 对象。

代码 4-25

```java
public class People {
    ……
    public People() {
        System.out.println("调用了无参构造创建 People 对象");
    }
```

```java
    public People(int id) {
        System.out.println("调用了有参构造创建 People 对象");
    }
}

public class Student extends People {
    public Student(){
        System.out.println("调用了无参构造创建 Student 对象");
    }
    public Student(int id){
        System.out.println("调用了有参构造创建 Student 对象");
    }
}

public class Test {
    public static void main(String[] args) {
        Student student = new Student();
        Student student2 = new Student(1);
    }
}
```

程序运行结果如图 4-15 所示。

图 4-15

可以看到，创建 Student 对象之前一定会优先创建 People 对象，同时无论是调用 Student 的有参构造还者是无参构造，创建 People 对象都是通过调用其无参构造来完成的。那么问题来了，在创建 People 对象时，会不会调用其他类的构造函数呢？即 People 类有没有自己的父类？答案是 People 也有自己的父类，只不过这个父类不是我们自定义的，而是 Java 提供的。Java 中的每个类都有一个共同父类 Object，Object 类就是所有 Java 类的根（老祖宗），所有的 Java 类都是由 Object 类派生出来的。

我们现在明白了在创建子类时会默认调用父类的无参构造来创建父类对象，那么能否让父类调用有参构造来创建对象呢？答案是可以的，如何完成？需要使用 super 关键字，super 关键字和 this 关键字类似，但用法完全不同，this 用作访问当前类的属性和方法，super 用作子类访问父类的属性和方法。要调用父类的有参构造，可以修改 Student 类如代码 4-26 所示。

代码 4-26

```
public class Student extends People {
    public Student(){
        super(1);
        System.out.println("调用了无参构造创建 Student 对象");
    }
    public Student(int id){
        super(1);
        System.out.println("调用了有参构造创建 Student 对象");
    }
}
```

再次运行测试类代码，结果如图 4-16 所示。

图 4-16

可以看到当前代码全部是通过有参构造来创建 People 对象，同理调用父类无参构造的代码是 super()，并且这种方式是默认设置。当我们手动在子类构造方法中做出修改时，会覆盖掉默认的方式，改为调用父类有参构造的方式。

好了，我们已经学习了如何在子类构造函数中调用父类构造函数，那么在子类普通方法中如何调用父类的属性和普通方法呢？同样是使用 super 关键字，例如访问属性："super.属性名"，调用普通方法："super.方法名();"。需要强调的是，子类只能访问父类的公有属性和方法，即使用 public 修饰的属性和方法，无法访问私有 private 修饰的属性和方法，父类的私有属性可以通过它的公有方法来访问和修改，具体调用如代码 4-27 所示。

代码 4-27

```
public class Student extends People {
    public void show(){
        super.setName("张三");
        System.out.println(super.getName());
    }
}

public class Test {
    public static void main(String[] args) {
        Student student = new Student();
        student.show();
    }
}
```

程序运行结果如图 4-17 所示。

图 4-17

4.4.3 子类访问权限

上一章我们讲了子类可以通过 super 关键字来访问父类的属性和方法。但不是所有的父类属性和方法都可以被子类访问，那么父类的哪些属性和方法是可以被子类访问的呢？在解答这个问题之前，我们首先要学习访问权限修饰符。访问权限修饰符可以用来修饰类、属性和方法，不同的访问权限修饰符表示不同的作用域，包括 public、protected、默认修饰符和 private 这 4 种修饰符。一般使用 public 来修饰类，我们这里主要说的是对属性和方法的访问权限修饰符，其作用域如表 4-1 所示。

表 4-1

	同一个类	同一个包	不同包	子类
public	可以访问	可以访问	可以访问	可以访问
protected	可以访问	可以访问	不可以访问	可以访问
默认修饰符	可以访问	可以访问	不可以访问	不可以访问
private	可以访问	不可以访问	不可以访问	不可以访问

通过表 4-1 可以看到，子类只能访问父类 public 和 protected 修饰的属性和方法，默认修饰符和 private 修饰的属性和方法不能访问。这里引入了一个新的概念：包（package）。为什么要有包呢？包用来管理 Java 类，类似于我们用不同的文件夹管理不同的文件，一个项目中不可避免地会出现同名的 Java 类，为了防止产生冲突，可以把同名的 Java 类分别放入不同的包中，如图 4-18 所示。

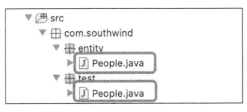

图 4-18

包的作用：(1) 管理 Java 类，便于查找和使用相应的文件；(2) 区分同名的类，防止命名冲突；(3) 实现访问权限控制。创建包之后，该包中的所有 Java 代码第一行必须添加包信息，使用 package 声明包，如图 4-19 所示。

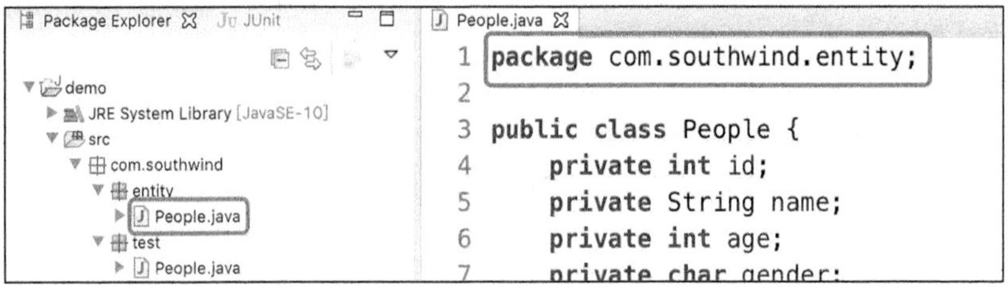

图 4-19

包的命名规范：包名由小写字母组成，不能以.开头或结尾。包名一般由小写字母组成，可以包含数字，但不能以数字开头，使用"."来分层，不能将"."用作开头或结尾，命名方式一般采用网络域名的反向输出，如 com.southwind.test 和 com.southwind.entity。在一个类中调用不同包的类时，需要使用 import 关键字导入该类，语法："import 包名.类名;"。Eclipse 会自动提示需要导类，如图 4-20 所示。

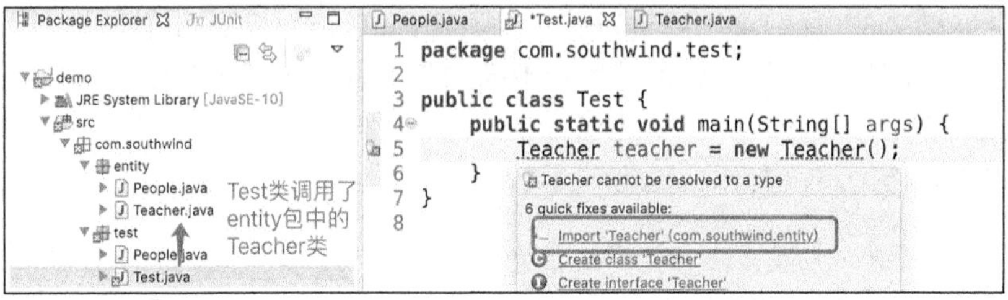

图 4-20

导入成功后如图 4-21 所示。

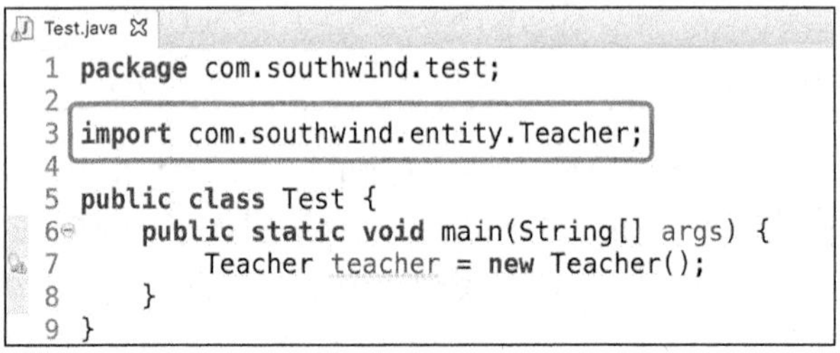

图 4-21

例子，下列代码的运行结果是什么呢？

```java
public class People {
    private int id = 3;
    public People(){
        System.out.println("编号是"+id);
    }
    public void setId(int id){
        this.id = id;
    }
    public void show(){
        System.out.println("编号是"+id);
```

```
    }
}

public class Student extends People {
    public Student(int id){
        super.setId(id);
    }
}

public class Test {
    public static void main(String[] args) {
        Student student = new Student(1);
        student.show();
    }
}
```

本题考察的知识点有两个：(1)创建子类对象时会默认调用父类无参构造；(2)子类调用父类的公有方法。所以 main 方法的第 1 行，"Student student = new Student(1);"调用了 Student 的有参构造，会默认调用 People 的无参构造，People 的无参构造函数会打印"编号是"和 id，id 是 People 的成员变量，值为 3，所以控制台输出为"编号是 3"。同时 Student 的有参构造调用了 People 的 setId()方法，参数为 1，所以 People 的成员变量 id 的值被修改为 1。接下来调用 student 从 People 继承来的 show()方法，会再次打印"编号是"和 id，此时控制台输出"编号是 1"，程序的运行结果如图 4-22 所示。

图 4-22

4.4.4 方法重写

子类在继承父类方法的基础上，对父类方法重新定义并覆盖的操作叫作方法重写。儿子继承了父亲的房子，但是对房子的装修风格不满意，于是把之前的装修成果全部拆掉，按照自己的审美重新装修，就类似于方法重写的概念。如代码 4-28 所示，在父类 People 中定义了 show()方法，打印"输出人员信息"，子类 Student 和 Teacher 继承了父类 People，同时继承了 show()方法，在测试类中创建 Student 和 Teacher 对象，调用 show()方法，看到结果打印了两次"输出人员信息"，此时并没有区分出 Student 和 Teacher，即没有体现出子类的特有信息。

代码 4-28

```
public class People {
```

```java
    public void show() {
        System.out.println("输出人员信息");
    }
}


public class Student extends People {

}


public class Teacher extends People {

}


public class Test {
    public static void main(String[] args) {
        Student student = new Student();
        student.show();
        Teacher teacher = new Teacher();
        teacher.show();
    }
}
```

程序运行结果如图 4-23 所示。

图 4-23

现在使用方法重写对代码进行优化，Student 和 Teacher 如代码 4-29 所示。

代码 4-29

```java
public class Student extends People {
    //方法重写
    @Override
    public void show() {
        // TODO Auto-generated method stub
        System.out.println("这是一个学生");
    }
}


public class Teacher extends People {
    //方法重写
    @Override
    public void show() {
        // TODO Auto-generated method stub
```

```
        System.out.println("这是一个老师");
    }
}
```

再次运行测试类，结果如图 4-24 所示。

图 4-24

通过重写的方式可以实现子类完成特定需求的功能,需要注意的是构造方法不能被重写。方法重写的规则：(1) 父子类的方法名相同；(2) 父子类的方法参数列表相同；(3) 子类方法返回值与父类方法返回值类型相同或者是其子类。(4) 子类方法的访问权限不能小于父类。

(1) 和 (2) 很好理解，重点说明 (3) 和 (4)。要求子类方法返回值与父类方法返回值类型相同或者是其子类，如代码 4-30 所示。

代码 4-30

```java
public class People {
    public Object getObj(){
        Object obj = new Object();
        return obj;
    }
}
//返回值相同的重写。

public class Student extends People {
    //方法重写
    public Object getObj(){
        Object obj = new Object();
        return obj;
    }
}
//子类方法返回值类型是父类方法返回值类型子类的重写。

public class Student extends People {
    //方法重写
    public String getObj(){
        return "这是一个学生";
    }
}
```

子类方法的访问权限不能小于父类。这个规则跟上一节我们讲过的访问权限修饰符有关，我们知道访问权限修饰符有 4 种，按照作用域范围从大到小排列为：public>protected>默认修饰符>private。若父类方法的访问权限修饰符为 public，则子类重写方法的访问权限修饰符只能是 public；若父类方法的访问权限修饰符为 protected，则子类重写方法的访问权限修饰符可以是 public 和 protected；若父类方法的访问权限修饰符为默认修饰符，则子类重写方法的访问权限修饰符可以是 public、protected、默认修饰符。父类的静态方法不能被子类重写为非静态方法，父类的非静态方法不能被子类重写为静态方法，父类的私有方法不能被子类重写。

4.4.5　方法重写 VS 方法重载

对于初学者来说，方法重写和方法重载很容易产生混淆，一张表带你了解两者的区别，如表 4-2 所示。

表 4-2

	所在位置	方法名	参数列表	返回值	访问权限
方法重写	子类	相同	相同	相同或是其子类	不能小于父类
方法重载	同一个类	相同	不同	没有要求	没有要求

4.5　多态

4.5.1　什么是多态

面向对象有三大特征：封装、继承和多态，在前面的章节中我们已经学习了封装和继承，本节我们来学习多态。多态的概念本身是比较抽象的，简单解释就是一个事物有多种表现形态，具体到 Java 程序中，就是定义一个方法，在具体的生产环境中根据不同的需求呈现出不同的业务逻辑，我们通过下面这个例子来理解什么是多态。业务场景：书店的普通会员买书，收银员为该会员计算优惠折扣。思路：先创建普通会员类，定义买书方法描述该会员买书的优惠折扣，再创建收银员类，将普通会员对象作为成员变量，在结算方法中调用普通会员的买书方法，如代码 4-31 所示。

代码 4-31

```java
public class OrdinaryMember {
    public void buyBook() {
        System.out.println("普通会员买书打 9 折");
    }
}


public class Cashier {
    private OrdinaryMember ordinaryMember;
    //getter、setter 方法

    public void settlement() {
        this.ordinaryMember.buyBook();
```

```
        }
    }

    public class Test {
        public static void main(String[] args) {
            OrdinaryMember ordinaryMember = new OrdinaryMember();
            Cashier cashier = new Cashier();
            cashier.setOrdinaryMember(ordinaryMember);
            cashier.settlement();
        }
    }
```

程序的运行结果如图4-25所示。

图4-25

用户希望买书能享受更大优惠，就把普通会员升级为超级会员，买书可以享受6折优惠。现在用超级会员再次购书，这时候就需要创建超级会员类，并且Cashier类和Test类也需要做相应的修改，如代码4-32所示。

代码4-32

```
    public class SuperMember {
        public void buyBook() {
            System.out.println("超级会员买书打6折");
        }
    }


    public class Cashier {
        private SuperMember superMember;
        //getter、setter方法

        public void settlement() {
            this.superMember.buyBook();
        }
    }


    public class Test {
        public static void main(String[] args) {
            SuperMember superMember = new SuperMember();
            Cashier cashier = new Cashier();
            cashier.setSuperMember(superMember);
            cashier.settlement();
```

```
        }
    }
```

程序运行结果如图 4-26 所示。

图 4-26

这种方式存在明显不足,当需求发生改变时需要频繁地修改代码,代码的扩展性、维护性较差。使用多态可以进行优化,多态的思路是:创建 Member 类,作为 OrdinaryMember 和 SuperMember 的父类,在 OrdinaryMember 和 SuperMember 中分别对父类方法进行重写,在 Cashier 类中定义 Member 类型的成员变量,如代码 4-33 所示。

代码 4-33

```java
public class Member {
    public void buyBook() {
    }
}


public class OrdinaryMember extends Member {
    //重写父类方法
    @Override
    public void buyBook() {
        // TODO Auto-generated method stub
        System.out.println("普通会员买书打 9 折");
    }
}


public class SuperMember extends Member{
    //重写父类方法
    @Override
    public void buyBook() {
        // TODO Auto-generated method stub
        System.out.println("超级会员买书打 6 折");
    }
}


public class Cashier {
    private Member member;
    //getter、setter 方法

    public void settlement() {
        this.member.buyBook();
    }
```

```java
}
public class Test {
    public static void main(String[] args) {
        Member member = new OrdinaryMember();
        Cashier cashier = new Cashier();
        cashier.setMember(member);
        cashier.settlement();
    }
}
```

同样的会员升级，从普通会员升级为超级会员，此时就不需要修改 Cashier 类，只需要在 Test 类的 main 方法中作出修改："Member member = new SuperMember();"，如代码 4-34 所示。

代码 4-34

```java
public class Test {
    public static void main(String[] args) {
        Member member = new SuperMember();
        Cashier cashier = new Cashier();
        cashier.setMember(member);
        cashier.settlement();
    }
}
```

如果业务需要扩展，想升级为其他类型的会员，只需要创建对应的会员类并继承 Member 类，然后在 main 方法中作出修改，将该会员的实例化对象赋给 member 即可。这就是多态，从 main 方法的角度来看这段代码，我们不去定义具体的 OrdinaryMember 或者 SuperMember，而是定义 Member，然后将具体的实例化对象赋给 Member，即同一个 Member 有多种表现形式。

4.5.2 多态的使用

代码 4-34 中有这样一行代码 "Member member = new SuperMember();"，它用于定义父类变量 member，然后将子类 SuperMember 的实例化对象赋值给 member，即父类引用指向子类对象，是多态的具体表现形式。在实际开发中，多态主要有两种表现形式：一种是定义方法时形参为父类，调用方法时传入的参数为子类对象；另一种是定义方法时返回值的数据类型为父类，调用方法时返回子类对象。

1. 定义方法时形参为父类，调用方法时传入的参数为子类对象，如代码 4-35 所示。

代码 4-35

```java
public class Cashier {
    public void settlement(Member member) {
        member.buyBook();
    }
}
```

```
public class Test {
    public static void main(String[] args) {
        OrdinaryMember ordinaryMember = new OrdinaryMember();
        SuperMember superMember = new SuperMember();
        Cashier cashier = new Cashier();
        cashier.settlement(ordinaryMember);
        cashier.settlement(superMember);
    }
}
```

运行结果如图 4-27 所示。

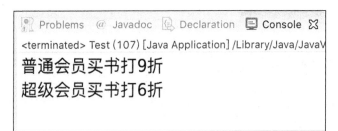

图 4-27

2. 定义方法时返回值的数据类型为父类，调用方法时返回子类对象，如代码 4-36 所示。

代码 4-36

```
public class Cashier {
    public Member getMember(String name) {
        if(name.equals("ordinaryMember")) {
            return new OrdinaryMember();
        }else {
            return new SuperMember();
        }
    }
}
```

无论是上述的哪种方式，成立的前提都是因为父类引用可以指向子类对象，即"Member member = new OrdinaryMember ();"或者"Member member = new SuperMember();"。也就是说我们可以把一个 OrdinaryMember 对象或者一个 SuperMember 对象当作 Member 对象来使用。在现实世界中也是合理的，比如张三是普通会员，我们说张三是会员也是没有错的。但是反过来就行不通了，如果我们说会员是张三，这句话从逻辑上来讲是行不通的。因为如果李四办理了超级会员，那么李四也是会员，同理王五也可以是会员，所以直接说会员是张三就以偏概全了。既然在现实世界中行不通，那么在程序的世界中也是行不通的，即我们不能把一个子类引用指向父类对象，如图 4-28 所示。

这是一种数据类型错误，例如不能把 Member 转为 OrdinaryMember。但是如果我们一定要将 Member 转为 OrdinaryMember 呢？我一定要说会员就是张三怎么办？好吧，既然你这么强势，那就随你吧！强制性地把 Member 转为 OrdinaryMember，这种方式叫作强制类型转换，需要在目标对象前加括号，括号内注明强制转换之后的数据类型，如代码 4-37 所示。

```
1  package com.southwind.test;
2
3  import com.southwind.entity.Cashier;
4  import com.southwind.entity.Member;
5  import com.southwind.entity.OrdinaryMember;
6
7  public class Test {
8      public static void main(String[] args) {
9          OrdinaryMember ordinaryMember = new Member();
10     }
11 }
12
```
Type mismatch: cannot convert from Member to OrdinaryMember
2 quick fixes available:
- Add cast to 'OrdinaryMember'
- Change type of 'ordinaryMember' to 'Member'

图 4-28

代码 4-37

```
public class Test {
    public static void main(String[] args) {
        OrdinaryMember ordinaryMember = (OrdinaryMember) new Member();
    }
}
```

综上所述，具有父子级关系的两个对象可以相互转换，子类转父类即父类引用指向子类对象，可以自动完成，无需强制转换。例如我们说张三是会员，这句话逻辑上是行得通的，同时也叫向上转型。父类转子类即子类引用指向父类对象，不能自动完成转换，需要强制转换。例如我们说会员是张三，以偏概全了，需要强制干预，这种方式也叫向下转型。

4.5.3 抽象方法和抽象类

回到前面章节中我们举过的例子：通过子类重写父类方法的形式实现多态，从而提高程序的扩展性，父类 Member 以及两个子类 OrdinaryMember 和 SuperMember 的定义如代码 4-38 所示。

代码 4-38

```
public class Member {
    public void buyBook(){
    }
}


public class OrdinaryMember extends Member {
    //重写父类方法
    @Override
    public void buyBook() {
        // TODO Auto-generated method stub
        System.out.println("普通会员买书打 9 折");
    }
}
```

```java
public class SuperMember extends Member{
    //重写父类方法
    @Override
    public void buyBook() {
        // TODO Auto-generated method stub
        System.out.println("超级会员买书打 6 折");
    }
}
```

观察上述 3 个类，OrdinaryMember 和 SuperMember 会分别对 Member 的 buyBook() 方法进行重写，也就是说无论 Member 的 buyBook()方法里写了什么，最后都会被子类所覆盖，所以 Member 的 buyBook()方法体就是无意义的。那么我们就可以只声明 buyBook() 方法，而不需要定义 buyBook()的方法体，这种没有方法体的方法叫作抽象方法。声明抽象方法时需要添加 abstract 关键字，如代码 4-39 所示。

代码 4-39

```java
public abstract void buyBook ();
```

一旦类中定义了抽象方法，则该类也必须声明为抽象类，需要在类定义处添加 abstract 关键字，如代码 4-40 所示。

代码 4-40

```java
public abstract class Member {
    public abstract void buyBook();
}
```

抽象类与普通类的区别是抽象类不能被实例化，抽象方法与普通方法的区别是抽象方法没有方法体。抽象类中可以没有抽象方法，但包含了抽象方法的类必须被定义为抽象类。即我们可以在抽象类中定义普通方法，但是在普通类中不能定义抽象方法，如图 4-29 和图 4-30 所示。

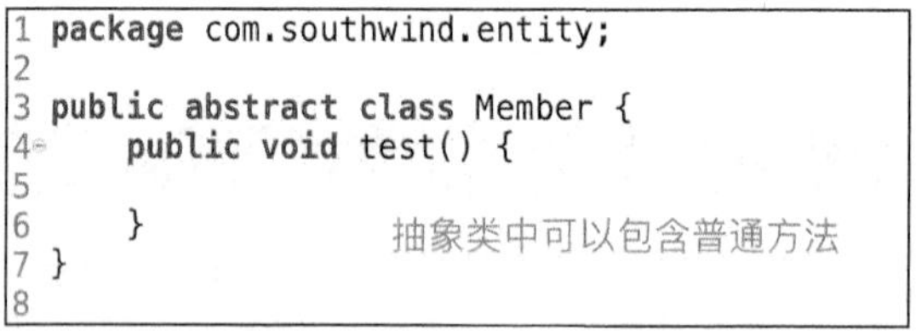

```
1 package com.southwind.entity;
2
3 public abstract class Member {
4     public void test() {
5
6     }                          抽象类中可以包含普通方法
7 }
8
```

图 4-29

```
1 package com.southwind.entity;
2
3 public class Member {
4     public abstract void buyBook();     普通类中不能
5 }                                        包含抽象方法
6
```

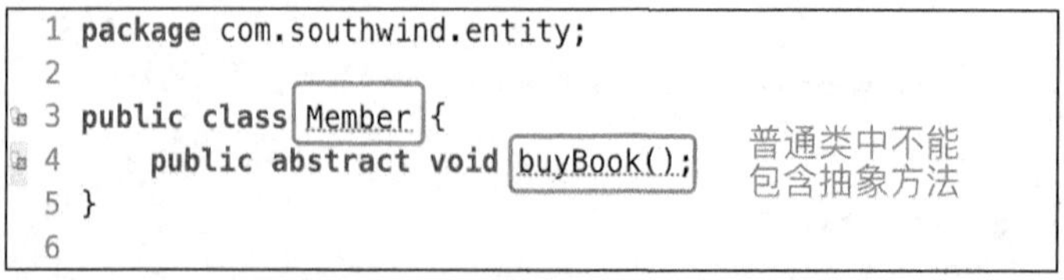

图 4-30

既然抽象类不能被实例化，抽象方法也没有方法体，那么为什么要创建抽象类和抽象方法呢？到底有什么用呢？抽象类和抽象方法需要结合多态来使用，我们知道构建多态的基础是类的继承和方法重写。既然有重写就意味着父类方法只需要声明，不需要具体的实现，具体实现由子类来完成，父类只是一个抽象的概念或者模板。那么我们就可以把父类定义为抽象类，需要被子类重写的方法定义为抽象方法，并针对这个抽象的概念进行编程。在具体执行时，给父类赋予不同的子类就会实现不同的功能，这就是多态的意义，即抽象类和抽象方法的作用。在实际开发中我们需要创建抽象类的子类来完成开发，如代码 4-41 所示。

代码 4-41

```java
public abstract class Member {
    public abstract void buyBook();
}


public class OrdinaryMember extends Member {
    //实现父类的抽象方法
    @Override
    public void buyBook() {
        // TODO Auto-generated method stub
        System.out.println("普通会员买书打 9 折");
    }
}


public class SuperMember extends Member{
    //实现父类的抽象方法
    @Override
    public void buyBook() {
        // TODO Auto-generated method stub
        System.out.println("超级会员买书打 6 折");
    }
}


public class Test {
    public static void main(String[] args) {
        Member member = new OrdinaryMember();
        member.buyBook();
        member = new SuperMember();
        member.buyBook();
    }
}
```

运行结果如图 4-31 所示。

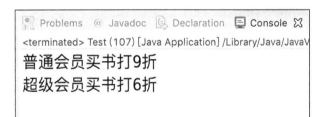

图 4-31

继承了抽象类的子类必须重写父类的抽象方法，以完成具体的方法实现，如图 4-32 所示。

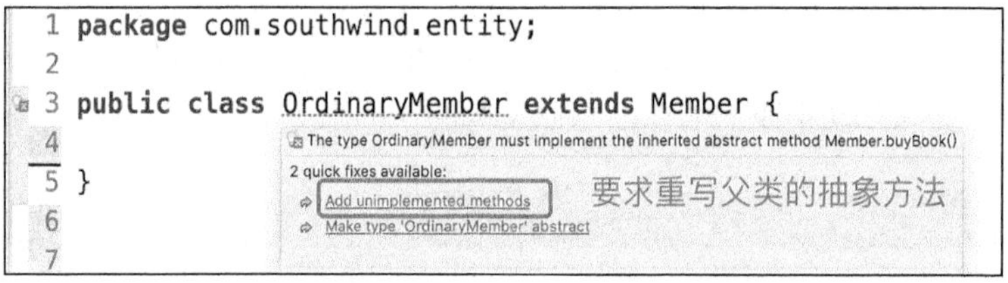

```
1  package com.southwind.entity;
2
3  public class OrdinaryMember extends Member {
     The type OrdinaryMember must implement the inherited abstract method Member.buyBook()
     2 quick fixes available:
       Add unimplemented methods            要求重写父类的抽象方法
       Make type 'OrdinaryMember' abstract
4
5  }
6
7
```

图 4-32

如果子类也是抽象类，可以不用重写父类的抽象方法，如图 4-33 所示。

```
1  package com.southwind.entity;
2
3  public abstract class OrdinaryMember extends Member {
4
5  }
```

图 4-33

4.6 小结

本章我们进入到 Java 的核心思想——面向对象编程部分。面向对象是一种编程思想，其核心是将程序中的所有参与角色都抽象成对象，然后通过对象之间的相互调用关系来完成需求。我们应该将学习的重点放在建立面向对象编程思想上，技术只是工具，编程思想才是核心竞争力。建立起自己的编程思想，即使让你使用另外一种编程语言进行开发，你也可以快速上手。

本章还介绍了 Java 是如何实现面向对象编程的，包括面向对象的三大特征：封装、继承和多态。这里需要注意，关于面向对象的特征也有 4 种的说法：封装、继承、多态和抽象，这种说法也是对的，而且抽象为实现多态提供了基础。

第 5 章　面向对象进阶

> 在前面的章节中我们学习了面向对象思想的基本概念，对面向对象的三大特征（封装、继承和多态）都做了详细的阐述，相信大家对这些概念已经有了一定的理解和掌握。面向对象更重要的是理解其编程思想，具备把程序模块化成对象的能力，思想的建立需要不断地思考，勤加练习，本章我们继续学习面向对象的高级部分。

5.1　Object 类

5.1.1　认识 Object 类

　　Java 是通过类来构建代码结构的，类分为两种：一种是 Java 提供的，无需开发者自定义，可直接调用；另外一种是由开发者根据不同的业务需求自定义的类。所以我们写的 Java 程序，其实就是由 Java 提供的类和自定义的类组成的，打开 Eclipse，在 JRE System Library 中存放的就是 Java 提供的类，开发者自定义的类存放在 src 目录下，如图 5-1 和图 5-2 所示。

图 5-1

JRE System Library 中的类全部是编译之后的字节码文件，即 class 格式的文件，我们可以看到源码，但是不能修改，如图 5-3 所示。

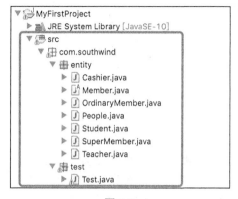

图 5-2

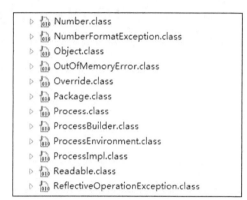

图 5-3

Object 就是 Java 提供的一个类，位于 java.lang 包中，该类是所有类的直接父类或间接父类。无论是 Java 提供的类，还是开发者自定义的类，都是 Object 的直接子类或间接子类。或者说 Java 中的每一个类都是 Object 的后代，Object 是所有类的祖先。一个类在定义时如果不通过 extends 指定其直接父类，系统就会自动为该类继承 Object 类，Object 类的源码如代码 5-1 所示。

代码 5-1

```java
public class Object {
private static native void registerNatives();
    static {
        registerNatives();
    }
    public final native Class<?> getClass();
    public native int hashCode();
    public boolean equals(Object obj) {
        return (this == obj);
    }
    protected native Object clone() throws CloneNotSupportedException;
    public String toString() {
        return getClass().getName() + "@" + Integer.toHexString(hashCode());
    }
    public final native void notify();
    public final native void notifyAll();
    public final native void wait(long timeout) throws InterruptedException;
    public final void wait(long timeout, int nanos) throws InterruptedException {
        if (timeout < 0) {
            throw new IllegalArgumentException("timeout value is negative");
        }
        if (nanos < 0 || nanos > 999999) {
            throw new IllegalArgumentException(
                        "nanosecond timeout value out of range");
        }
        if (nanos > 0) {
            timeout++;
        }
```

```
        wait(timeout);
    }
    public final void wait() throws InterruptedException {
        wait(0);
    }
    protected void finalize() throws Throwable { }
}
```

可以看到 Object 类中提供了很多用 public 和 protected 修饰的方法，子类是可以直接继承这些方法的，即 Java 中的任何一个类，都可以调用 Object 类中的 public 和 protected 方法，当然 private 是不能调用的，如图 5-4 所示。

```
package com.southwind.entity;

public class People {
    public void test() throws Throwable {
        hashCode();
        getClass();
        equals(null);
        clone();
        toString();
        notify();
        notifyAll();
        wait();
        wait(100L);
        wait(100L, 100);
    }
}
```
继承Object类的方法

图 5-4

5.1.2 重写 Object 类的方法

上一节我们介绍了 Object 是所有类的父类，每一个类都可以直接继承 Object 类的非私有方法，实例化对象可以直接调用这些方法。但是通常情况下不会直接调用这些方法，而是需要对它们进行重写，因为父类中统一的方法并不能适用于所有的子类。就像老爹房子的装修风格是老爹喜欢的，儿子们审美各有不同，老爹的房子并不能满足他们的需求，所以儿子们会把房子的旧装修覆盖掉，重新装修以适应他们的需求。这种方式是多态的一种体现，父类信息通过不同的子类呈现出不同的形态，接下来我们就一起看看 Object 类经常被子类所重写的那些方法，如表 5-1 所示。

表 5-1

方法	描述
public String toString()	以字符串的形式返回该类的实例化对象信息
public boolean equals(Object obj)	判断两个对象是否相等
public native int hashCode()	返回对象的散列码

先来看看这 3 个方法的具体实现。

toString() 方法的实现如图 5-5 所示。

```
 * @return  a string representation of the object.
 */
public String toString() {
    return getClass().getName() + "@" + Integer.toHexString(hashCode());
}
```

图 5-5

原生的toString()方法会返回对象的类名以及散列值,直接打印对象默认调用toString()方法,如代码5-2所示。

代码 5-2

```java
public class Test {
    public static void main(String[] args) {
        People people = new People();
        people.setId(1);
        people.setName("张三");
        people.setAge(22);
        people.setGender('男');
        System.out.println(people);
    }
}
```

程序的运行结果如图5-6所示。

图 5-6

但是在实际开发中返回这样的信息意义不大,我们更希望看到的是对象的属性值,而非它的内存地址,所以我们需要对toString()方法进行重写,如代码5-3所示。

代码 5-3

```java
public class People {
    ……
    @Override
    public String toString() {
        return "People [id=" + id + ", name=" + name + ", age=" + age + ", gender=" + gender + "]";
    }
}

public class Test {
    public static void main(String[] args) {
        People people = new People();
        people.setId(1);
        people.setName("张三");
        people.setAge(22);
        people.setGender('男');
        System.out.println(people);
    }
}
```

程序的运行结果如图5-7所示。

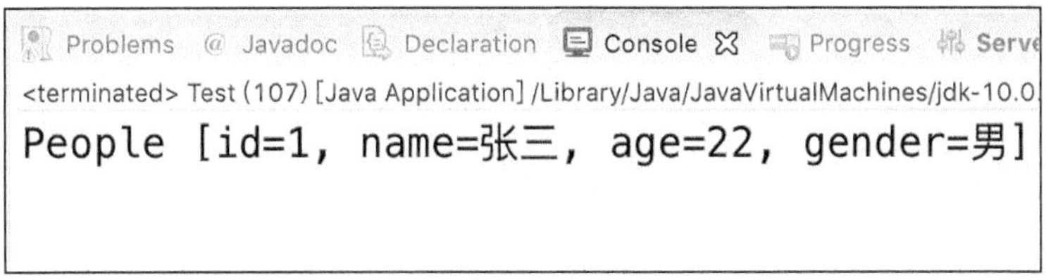

```
 Problems  @ Javadoc  Declaration  Console ⊠  Progress  Serve
<terminated> Test (107) [Java Application] /Library/Java/JavaVirtualMachines/jdk-10.0.
People [id=1, name=张三, age=22, gender=男]
```

图 5-7

equals()方法的实现如图 5-8 所示。

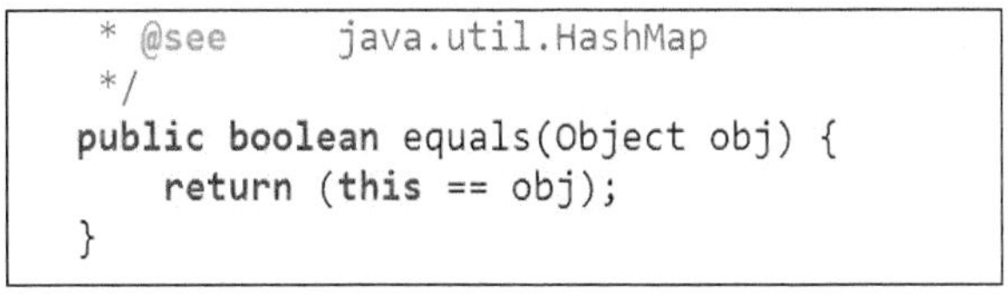

```
 * @see         java.util.HashMap
 */
public boolean equals(Object obj) {
    return (this == obj);
}
```

图 5-8

通过内存地址对两个对象进行判断,即两个对象的引用必须指向同一块内存程序才会认为它们相等,但是在不同的场景下,这种方式不见得都适用。比如两个字符串"String str1 = new String("Hello");" 和 "String str2 = new String("Hello");",虽然 str1 和 str2 是两个完全不同的对象,但是它们的值是相等的,就可以认为这两个字符串相等。我们需要对 equals()方法进行重写,String 类已经完成了重写的工作,直接使用即可,重写的代码如代码 5-4 所示。

代码 5-4

```java
public boolean equals(Object anObject) {
    if (this == anObject) {
        return true;
    }
    if (anObject instanceof String) {
        String anotherString = (String)anObject;
        int n = value.length;
        if (n == anotherString.value.length) {
            char v1[] = value;
            char v2[] = anotherString.value;
            int i = 0;
            while (n-- != 0) {
                if (v1[i] != v2[i])
                    return false;
                i++;
            }
            return true;
        }
    }
    return false;
}
```

你可以看到 String 类中对 equals()方法的重写,是将两个字符串中的每一个字符依次取出进行比对,如果所有字符完全相等,则认为两个对象相等,否则不相等,字符串比较的过程如代码 5-5 所示。

代码 5-5

```java
public class Test {
    public static void main(String[] args) {
        String str1 = new String("Hello");
        String str2 = new String("Hello");
        System.out.println(str1.equals(str2));
    }
}
```

程序的运行结果如图 5-9 所示。

图 5-9

自定义类也可以根据需求对 equals()方法进行重写，如我们定义一个 People 类，创建该类的实例化对象，认为只要成员变量的值都相等就是同一个人，用程序的语言来表述就是两个对象相等，但是如果直接调用 equals()方法进行比较，结果却并不是我们所预期的，如代码 5-6 所示。

代码 5-6

```java
public class People {
    private int id;
    private String name;
    private int age;
    private char gender;
    //getter、setter 方法
}


public class Test {
    public static void main(String[] args) {
        People people = new People();
        people.setId(1);
        people.setName("张三");
        people.setAge(22);
        people.setGender('男');
        People people2 = new People();
        people2.setId(1);
        people2.setName("张三");
        people2.setAge(22);
        people2.setGender('男');
        System.out.println(people.equals(people2));
    }
}
```

程序的运行结果如图 5-10 所示。

图 5-10

现在对 People 类继承自 Object 类的 equals()方法进行重写,如果两个对象的成员变量值都相等,则它们就是同一个对象,具体实现如代码 5-7 所示。

代码 5-7

```
public class People {
    @Override
    public boolean equals(Object obj) {
        // TODO Auto-generated method stub
        People people = (People) obj;
        if(people.getId() == this.id && people.getName().equals(this.name) && people.getGender() == this.gender && people.getAge() == this.age){
            return true;
        }
        return false;
    }
}
```

再次运行程序,结果如图 5-11 所示。

图 5-11

hashCode()方法如图 5-12 所示。该方法返回一个对象的散列值,这个值是由对象的内存地址结合对象内部信息得出的,任何两个对象的内存地址肯定是不一样的。但是在上面举的例子中,我们认为如果两个 People 对象的成员变量值都相等,就是同一个对象,那么它们的散列值也应该相等,如果直接调用父类的 hashCode()方法,两个对象的散列值是不相等的,如代码 5-8 所示。

```
 * @return  a hash code value for this object.
 * @see     java.lang.Object#equals(java.lang.Object)
 * @see     java.lang.System#identityHashCode
 */
public native int hashCode();
```

图 5-12

代码 5-8

```java
public class Test {
    public static void main(String[] args) {
        People people = new People();
        people.setId(1);
        people.setName("张三");
        people.setAge(22);
        people.setGender('男');
        People people2 = new People();
        people2.setId(1);
        people2.setName("张三");
        people2.setAge(22);
        people2.setGender('男');
        System.out.println(people.hashCode());
        System.out.println(people2.hashCode());
    }
}
```

程序的运行结果如图 5-13 所示。

图 5-13

现在对 People 的 hashCode()方法进行重写，将 id*name*age*gender 的值作为结果返回，name 是字符串类型，值相等散列值就相等，具体实现如代码 5-9 所示。

代码 5-9

```java
public class People {
    ……
    @Override
    public int hashCode() {
        return this.id*this.name.hashCode()*this.age*this.gender;
    }
}
```

再次运行程序，结果如图 5-14 所示。

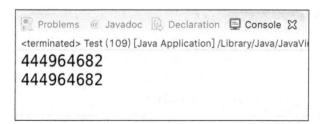

图 5-14

如此一来，成员变量值都相等的两个 People 对象，散列值也是相等的。

5.2 包装类

5.2.1 什么是包装类

Java 中的数据类型从本质上看可以分为两类：8 种基本数据类型和引用类型。8 种基本数据类型很好区分，那么什么是引用类型呢？这里大家可简单记住一句话，通过构造函数 new 出来的对象就是引用类型。基本类型的数据不是对象，引用类型的数据才能称之为对象。我们知道 Java 是面向对象的编程语言，某些场景需要用对象来描述基本类型的数据，如何来实现呢？就是通过包装类来完成的，包装类是 Java 提供的一组类，专门用来创建 8 种基本数据类型对应的对象。所以我们很容易得出一个结论：包装类一共有 8 个，这些类都保存在 java.lang 包中，基本数据类型对应的包装类如表 5-2 所示。

表 5-2

基本数据类型	包装类
byte	Byte
short	Short
int	Integer
long	Long
float	Float
double	Double
char	Character
boolean	Boolean

包装类的体系结构如图 5-15 所示。

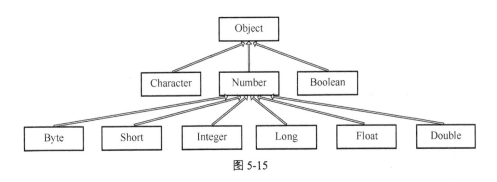

图 5-15

5.2.2　装箱与拆箱

装箱与拆箱是包装类的特有名词，装箱是指将基本数据类型转为对应的包装类对象，拆箱指将包装类对象转为对应的基本数据类型。

1．装箱

（1）public Type(type value)

每个包装类都提供了一个有参构造函数：public Type(type value)，用来实例化包装类对象，具体使用如代码 5-10 所示。

代码 5-10

```
public class Test {
   public static void main(String[] args) {
      byte b = 1;
      Byte byt = new Byte(b);
      short s = 2;
      Short shor = new Short(s);
      int i = 3;
      Integer integer = new Integer(i);
      long l = 4;
      Long lon = new Long(l);
      float f = 5.5f;
      Float flo = new Float(f);
      double d = 6.6;
      Double dou = new Double(d);
      char cha = 'J';
      Character charac = new Character(cha);
      boolean bo = true;
      Boolean bool = new Boolean(bo);
   }
}
```

（2）public Type(String value)/ public Type(char value)

每个包装类还有一个重载构造函数，Character 类的重载构造函数为：public Type(char value)；除 Character 类以外的其他的包装类重载构造函数为：public Type(String value)。具体使用如代码 5-11 所示。

代码 5-11

```
public class Test {
   public static void main(String[] args) {
      Byte byt = new Byte("1");
      Short shor = new Short("2");
      Integer integer = new Integer("3");
      Long lon = new Long("4");
      Float flo = new Float("5.5f");
      Double dou = new Double("6.6");
      Character charac = new Character('J');
      Boolean bool = new Boolean("true");
   }
}
```

需要注意的是，在 Boolean 类的构造函数中，当参数为 true 时，Boolean 值为 true；

当参数不为 true 时，Boolean 值都为 false。

（3）valueOf(type value)

每一个包装类都有一个 valueOf(type value)方法：public static Type valueOf(type value)，可以将基本数据类型转为包装类类型，具体使用如代码 5-12 所示。

代码 5-12

```java
public class Test {
    public static void main(String[] args) {
        byte b = 1;
        Byte byt = Byte.valueOf(b);
        short s = 2;
        Short shor = Short.valueOf(s);
        int i = 3;
        Integer integer = Integer.valueOf(i);
        long l = 4;
        Long lon = Long.valueOf(l);
        float f = 5.5f;
        Float flo = Float.valueOf(f);
        double d = 6.6;
        Double dou = Double.valueOf(d);
        char cha = 'J';
        Character charac = Character.valueOf(cha);
        boolean bo = true;
        Boolean bool = Boolean.valueOf(bo);
    }
}
```

（4）valueOf(String value)/valueOf(char value)

Character 有一个重载方法：public static Type valueOf(char value)，可以将 char 类型转为包装类类型。除 Character 类以外的其他包装类的重载方法：public static Type valueOf(String value)，可以将 String 类型转为包装类类型，具体使用如代码 5-13 所示。

代码 5-13

```java
public class Test {
    public static void main(String[] args) {
        Byte byt = Byte.valueOf("1");
        Short shor = Short.valueOf("2");
        Integer integer = Integer.valueOf("3");
        Long lon = Long.valueOf("4");
        Float flo = Float.valueOf("5.5f");
        Double dou = Double.valueOf("6.6");
        Character charac = Character.valueOf('J');
        Boolean bool = Boolean.valueOf("true");
    }
}
```

需要注意的是在 Boolean.valueOf(String value)中，当 value 为 true 时，Boolean 值为 true；当 value 不为 true 时，Boolean 值都为 false。

2．拆箱

（1）*Value()

每个包装类都有一个*Value()方法，*是包装类对应的基本数据类型的名称，通过该方法可以将包装类转为基本数据类型，具体使用如代码 5-14 所示。

代码 5-14

```java
public class Test {
    public static void main(String[] args) {
        Byte byt = Byte.valueOf("1");
        byte b = byt.byteValue();
        Short shor = Short.valueOf("2");
        short s = shor.shortValue();
        Integer integer = Integer.valueOf("3");
        int i = integer.intValue();
        Long lon = Long.valueOf("4");
        long l = lon.longValue();
        Float flo = Float.valueOf("5.5f");
        float f = flo.floatValue();
        Double dou = Double.valueOf("6.6");
        double d = dou.doubleValue();
        Character charac = Character.valueOf('J');
        char cha = charac.charValue();
        Boolean bool = Boolean.valueOf("true");
        boolean bo = bool.booleanValue();
    }
}
```

（2）parse*(String value)

除 Character 类以外的每一个包装类都有一个静态方法可以将字符串类型转为基本数据类型，具体使用如代码 5-15 所示。

代码 5-15

```java
public class Test {
    public static void main(String[] args) {
        byte b = Byte.parseByte("1");
        short s = Short.parseShort("2");
        int i = Integer.parseInt("3");
        long l = Long.parseLong("4");
        float f = Float.parseFloat("5.5f");
        double d = Double.parseDouble("6.6");
        boolean bo = Boolean.parseBoolean("true");
    }
}
```

需要注意的是，在 Boolean.parseBoolean (String value)中，当 value 为 true 时，Boolean 值为 true；当 value 不为 true 时，Boolean 值都为 false。

3．其他方法

toString(type value)：每一个包装类都有一个静态方法 toString(type value)，可以将基本数据类型转为 String 类型，具体使用如代码 5-16 所示。

代码 5-16

```java
public class Test {
```

```java
    public static void main(String[] args) {
        byte b = 1;
        String bstr = Byte.toString(b);
        short s = 2;
        String sstr = Short.toString(s);
        String i = Integer.toString(3);
        long l = Long.parseLong("4");
        String lstr = Long.toString(l);
        float f = Float.parseFloat("5.5f");
        String fstr = Float.toString(f);
        double d = Double.parseDouble("6.6");
        String dstr = Double.toString(d);
        boolean bo = Boolean.parseBoolean("true");
        String bostr = Boolean.toString(bo);
        String chastr = Character.toString('J');
    }
}
```

5.3 接口

5.3.1 什么是接口

接口是 Java 程序开发中很重要的一种思想，准确地讲不仅仅是 Java 编程，对于其他高级编程语言来说接口也是非常重要的，在实际开发中使用非常广泛。接口是由抽象类衍生出来的一个概念，并由此产生了一种编程方式：面向接口编程。我们已经掌握了面向对象编程思想，那什么是面向接口编程呢？面向接口编程不是一种思想，更准确地讲它应该是一种编程方式。面向接口编程就是将程序的业务逻辑进行分离，以接口的形式去对接不同的业务模块。接口只串联不实现，真正的业务逻辑实现交给接口的实现类来完成。当用户需求变更的时候，只需要切换不同的实现类，而不需要修改串联模块的接口，减少对系统的影响。

上面这段解释不是很容易理解，我们通过一个现实生活中的例子来类比面向接口编程，我们都知道计算机可以通过 USB 插口来连接外部设备，例如鼠标、U 盘、散热架等。假如没有 USB 插口这种配置，那么外部设备就是固定在计算机上的，不能更换。例如一个鼠标固定在计算机上，但我现在希望换成 U 盘，怎么办呢？我们可能就需要把计算机拆了，重新组装计算机和 U 盘的内部连接。同理，每一次希望更换外置设备的时候，都需要把计算机拆了，移除之前的结构并重新组装，这种方式很显然是不可取的。维护成本太高，效率太低，用专业的语言来描述，叫作耦合性太高，模块和模块结合得太紧密，不灵活，要更换就必须重新构建内部组成来替换原有的结构。

但是有了 USB 插口之后，上述问题就迎刃而解了。通过 USB 插口连接到电脑上的外部设备是非常灵活的，即插即用，需要更换就拔掉旧的，把新的设备插入即可，这就是接口的思想。在设计制作计算机的时候，不需要考虑到底是跟鼠标、U 盘还是散热架连接，只需要把这个对接的部分提取出来，设计成一个接口。计算机内部只需要跟这个接口进行连接即可，你插入鼠标，计算机就识别鼠标，插入 U 盘就识别 U 盘。USB 插口就相当于接口，鼠标、U 盘、散热架就相当于接口的实现类，这样就很好理解了。

使用 USB 插口的方式可以很好地实现计算机与外部设备之间的各种切换，这是因为它们的连接是松散的、不紧密的，用专业的语言来讲就叫作低耦合，模块和模块之间连接松散，自然很容易切换。这里我们又引入来耦合性这个概念，很显然在实际开发中应该尽量降低程序的耦合性，以提高程序的扩展性，便于维护。面向接口编程就具备以下优点：

- 能够最大限度地解耦，降低程序的耦合性；
- 使程序易于扩展；
- 有利于程序的后期维护。

5.3.2 如何使用接口

了解完接口的概念，以及我们为什么要使用接口，接下来就是如何使用接口了。接口在 Java 中是独立存在的一种结构，和类相似，我们需要创建一个接口文件，Java 中用 class 关键字来标识类，用 interface 来标识接口，基本语法如下：

```
public interface 接口名{
    public 返回值 方法名(参数列表);
}
```

看到定义接口的基本语法，你有没有一种很熟悉的感觉？没错，接口与抽象类非常相似，同样是定义了没有实现的方法，只是一个抽象的概念，没有具体实现。接口其实就是一个极度抽象的抽象类。为什么这么说呢？抽象类的概念我们知道，一个类中一旦存在没有具体实现的抽象方法，那么该类就必须定义为抽象类，同时抽象类中是允许存在非抽象方法的。但是接口完全不同，接口中不能存在非抽象方法，必须全部是抽象方法，所以我们说接口是极度抽象的抽象类。因为接口中全部是抽象方法，所以修饰抽象方法的 abstract 可以省略，不需要添加在方法定义处。当然，在定义方法时添加 abstract，程序也不会报错。

我们知道了接口中的方法必须全部是抽象的，那么接口中可以存在成员变量吗？答案是肯定的，接口中可以定义成员变量，但是有如下要求：

- 不能定义 private 和 protected 修饰的成员变量，只能定义 public 和默认访问权限修饰的成员变量。
- 接口中的成员变量在定义时必须被初始化。
- 接口中的成员变量都是静态常量，即可以直接通过接口访问，同时值不能被修改。

关于接口的创建，如代码 5-17 所示。

代码 5-17

```
public interface MyInterface {
    public int ID = 0;
    String NAME = "";
    public void test();
}
```

接口定义完成，接下来如何使用呢？对于类的使用我们已经很熟悉了，首先需要实例化一个类的对象，然后通过对对象的操作来完成功能。但是接口的使用就大不一样了，因为接口是不能被实例化的，它描述的是一个抽象的信息，抽象的信息当然是没有实例的。我们需要实例化的是接口的实现类。实现类就是对接口的抽象方法进行具体实现的，实现类本身就是一个普通的 Java 类，创建实现类的基本语法如下：

```
public class 实现类名 implements 接口名{
    public 返回值 方法名(参数列表){
}
}
```

通过关键字 implements 来指定实现类具体要实现的接口，在实现类的内部需要对接口的所有抽象方法进行实现，同时要求访问权限修饰符、返回值类型、方法名和参数列表必须完全一致，如代码 5-18 所示。

代码 5-18

```
public class MyImplements implements MyInterface{
    @Override
    public void test() {
        // TODO Auto-generated method stub
        System.out.println("实现了接口的抽象方法");
    }
}
```

在这里接口与继承也有一个可以对比的地方，继承只能实现单继承，即一个子类只能继承一个父类。接口可以多实现，即一个实现类可以同时实现多个接口。如何去理解这句话呢？接口实际上描述的是功能，让某个类实现接口，就是让该类具备某些功能。类比现实生活中的例子，一个孩子只能有一个亲爹，所以是单继承，但是一个孩子可以同时具备多个技能，所以是多实现。如代码 5-19 所示。

代码 5-19

```
public interface MyInterface {
    public void fly();
}

public interface MyInterface2 {
    public void run();
}

public class MyImplements implements MyInterface,MyInterface2{
    @Override
    public void fly() {
        // TODO Auto-generated method stub
        System.out.println("实现了 fly 的功能");
    }
    @Override
    public void run() {
        // TODO Auto-generated method stub
        System.out.println("实现了 run 的功能");
    }
```

```
    }

public class Test {
    public static void main(String[] args) {
        MyImplements myImplements = new MyImplements();
        myImplements.fly();
        myImplements.run();
    }
}
```

运行结果如图 5-16 所示。

图 5-16

5.3.3 面向接口编程的实际应用

面向接口编程是一种常用的编程方式，可以有效地提高代码的复用性，增强程序的扩展性和维护性，我们通过下面这个例子来学习什么是面向接口编程。

某工厂生产成品 A，主要由设备 A 来完成生产，用程序模拟这一过程。分别创建 Factory 类和 EquipmentA 类，并将 EquipmentA 设置为 Factory 的成员变量，在 Factory 的业务方法中调用 EquipmentA 的方法来完成生产，具体实现如代码 5-20 所示。

代码 5-20

1. 定义 EquipmentA 类

```
public class EquipmentA {
    public void work() {
        System.out.println("设备A运行，生产成品A");
    }
}
```

2. 定义 Factory 类

```
public class Factory {
    private EquipmentA equipmentA;
    //getter、setter方法

    public void work() {
        System.out.println("开始生产...");
        this.equipmentA.work();
    }
}
```

3. Test 类中的工厂开始生产成品

```
public class Test {
   public static void main(String[] args) {
      EquipmentA equipmentA = new EquipmentA();
      Factory factory = new Factory();
      factory.setEquipmentA(equipmentA);
      factory.work();
   }
}
```

运行结果如图 5-17 所示。

图 5-17

现在工厂接了一份新订单，要求生产成品 B，需要设备 B 来完成生产，用程序实现这一过程，首先需要创建 EquipmentB 类，同时修改 Factory 内部的属性，如代码 5-21 所示。

代码 5-21

1. 定义 EquipmentB 类

```
public class EquipmentB {
   public void work() {
      System.out.println("设备 B 运行，生产成品 B");
   }
}
```

2. 修改 Factory 类，将成员变量的数据类型改为 EquipmentB

```
public class Factory {
   private EquipmentB equipmentB;
   //getter、setter 方法

   public void work() {
      System.out.println("开始生产...");
      this.equipmentB.work();
   }
}
```

3. 在 Test 类中完成生产

```
public class Test {
   public static void main(String[] args) {
      EquipmentB equipmentB = new EquipmentB();
      Factory factory = new Factory();
      factory.setEquipmentB(equipmentB);
      factory.work();
   }
}
```

运行结果如图 5-18 所示。

图 5-18

这种方式需要修改 Factory 类的内部结构,如果此时需求又改回到生产成品 A 或者生产成品 C,就需要创建新的类,同时再次修改 Factory 类的属性信息,这种当需求发生变更就要频繁修改类结构的方式是我们应该避免的。这种结构的程序扩展性非常差,如何改进呢?使用面向接口编程即可。

分析:如何使用面向接口编程来优化程序呢?将 Equipment 以接口的形式集成到 Factory 类中,具体实现如代码 5-22 所示。

代码 5-22

1. 定义 Equipment 接口

```
public interface Equipment {
   public void work();
}
```

2. 定义 Equipment 接口的实现类 EquipmentA 和 EquipmentB

```
public class EquipmentA implements Equipment{
   @Override
   public void work() {
      // TODO Auto-generated method stub
      System.out.println("设备 A 运行,生产成品 A");
   }
}


public class EquipmentB implements Equipment{
   @Override
   public void work() {
      // TODO Auto-generated method stub
      System.out.println("设备 B 运行,生产成品 B");
   }
}
```

3. 修改 Factory 类,将 Equipment 接口设置为成员变量

```
public class Factory {
   private Equipment equipment;
   //getter、setter 方法
   
   public void work() {
```

```
        System.out.println("开始生产...");
        this.equipment.work();
    }
}
```

4. Test 类中的工厂生产成品 A

```
public class Test {
    public static void main(String[] args) {
        Equipment equipment = new EquipmentA();
        Factory factory = new Factory();
        factory.setEquipment(equipment);
        factory.work();
    }
}
```

运行结果如图 5-19 所示。

图 5-19

如果此时需要生产成品 B，修改起来就方便多了。因为嵌入到 Factory 类中的是接口，所以 Factory 类不需要修改，只需要将 EquipmentB 组件以接口的形式赋给 Factory 实例对象即可，Test 类中修改一处代码，如代码 5-23 所示。

代码 5-23

```
public class Test {
    public static void main(String[] args) {
        Equipment equipment = new EquipmentB();
        Factory factory = new Factory();
        factory.setEquipment(equipment);
        factory.work();
    }
}
```

运行结果如图 5-20 所示。

图 5-20

5.4 异常

5.4.1 什么是异常

Java 中的错误我们大致可以分为两类，一类是编译时错误，一般指语法错误；另一类是运行时错误。我们知道 Java 程序运行首先需要进行编译，将 Java 文件编译成计算机能够识别的字节码文件，这类错误在程序编译时就会暴露出来,会导致程序编译失败。IDE 集成开发环境都会对这种错误进行提示，即我们在编写代码时能看到的语法错误，就叫编译时错误。因为可以即时看到，所以这种错误一般都能避免。而运行时错误在我们编写代码的过程中以及程序编译期间都难以发现，甚至可以正常编译通过，但一旦运行就会报错，这类错误一般不容易发现。编写代码的过程中往往会因为疏忽导致运行时错误的出现，例如数组下标越界，把 0 当作除数等。

Java 是一门面向对象的编程语言，世间万物都可以看作对象，那么同理错误也可以看作一个对象。Java 中有一组类专门来描述各种不同的运行时错误，叫作异常类，Java 结合异常类提供了处理错误的机制，具体步骤就是当程序出现错误时，会创建一个包含错误信息的异常类的实例化对象，并将该对象提交给系统，由系统转交给能处理该异常的代码进行处理。异常分为两类，包括 Error 和 Exception，Error 指系统错误，由 Java 虚拟机生成，我们编写的程序无法处理。Exception 指程序运行期间出现的错误，我们编写的程序可以对其进行处理。

举个生活中的例子来类比 Error 和 Exception。当你骑自行车出去玩时，如果半路自行车链条掉了，我们把链条重新安装就可以继续骑行，这属于自己可以处理的问题，就是 Exception。如果前方的路塌陷了，所有车都无法通过了，这属于我们无法处理的问题，就是 Error。

5.4.2 异常的使用

异常的使用需要用到两个关键字 try 和 catch，并且这两个关键字需要结合起来使用，用 try 来监听可能会抛出异常的代码，一旦捕获到异常，生成异常对象并交给 catch 来处理，基本语法如下：

```
try{
//可能抛出异常
}catch(异常对象){
//处理异常
}
```

具体实现如代码 5-24 所示。

代码 5-24

```
public class Test {
    public static void main(String[] args) {
        try {
```

```
            int num = 10/0;
        }catch (Exception e) {
            // TODO: handle exception
            e.printStackTrace();
        }
    }
}
```

运行结果如图 5-21 所示。

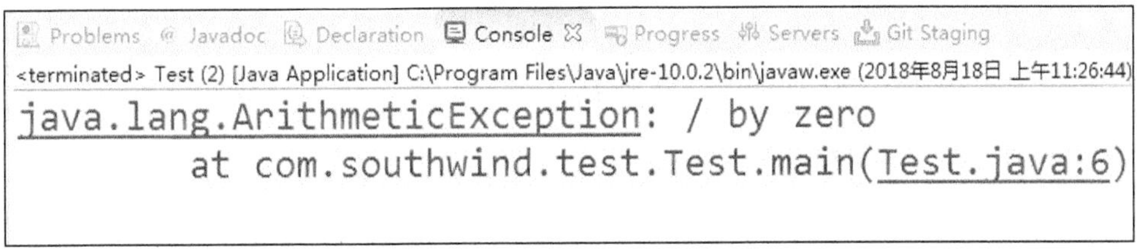

```
java.lang.ArithmeticException: / by zero
        at com.southwind.test.Test.main(Test.java:6)
```

图 5-21

可以看到因为 "int num = 10/0;" 代码中将 0 作为除数，所以程序在执行时会产生错误并自动生成一个 Exception 对象，在 catch 代码块中捕获 Exception 对象并进行处理，将错误信息打印出来。如果我们将代码修改为 "int num = 10/10;" 再次运行，就不会看到异常信息了。因为此时没有发生错误，就不会产生 Exception 对象，catch 代码块不执行。以上代码是异常最基本的使用，通常除了使用 try 和 catch 关键字，我们还需要用到 finally 关键字，这个关键字有什么作用呢？无论程序是否抛出异常，finally 代码块中的程序都会执行。finally 一般跟在 catch 代码块后面，基本语法如下：

```
try{
//可能抛出异常
}catch(异常对象){
//处理异常
}finally{
//必须执行的代码
}
```

具体实现如代码 5-25 所示。

代码 5-25

```
public class Test {
    public static void main(String[] args) {
        try {
            int num = 10/0;
        }catch (Exception e) {
            // TODO: handle exception
            e.printStackTrace();
        }finally {
            System.out.println("finally...");
        }
    }
}
```

运行结果如图 5-22 所示。

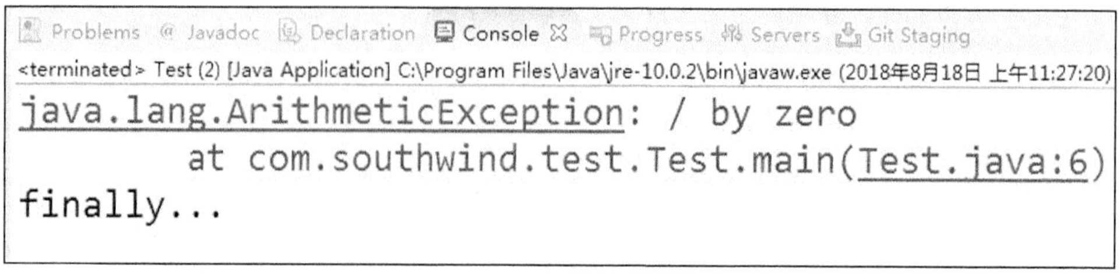

```
java.lang.ArithmeticException: / by zero
        at com.southwind.test.Test.main(Test.java:6)
finally...
```

图 5-22

现在我们对代码进行修改，不使用 finally，而在 catch 代码块后面直接执行 "System.out.println("finally...");"，如代码 5-26 所示。

代码 5-26

```java
public class Test {
    public static void main(String[] args) {
        try {
            int num = 10/0;
        }catch (Exception e) {
            // TODO: handle exception
            e.printStackTrace();
        }
        System.out.println("finally...");
    }
}
```

运行结果如图 5-23 所示。

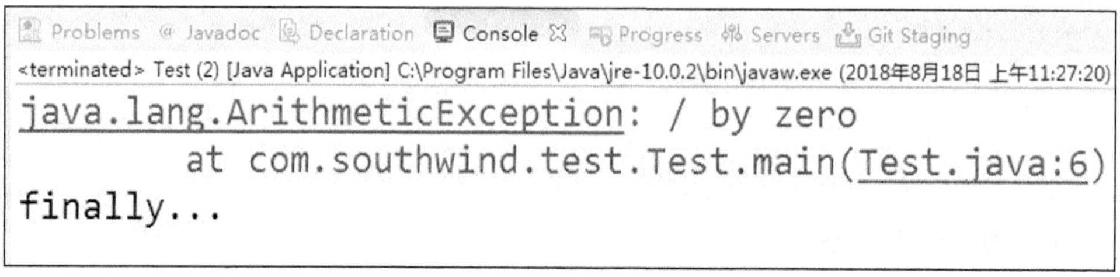

```
java.lang.ArithmeticException: / by zero
        at com.southwind.test.Test.main(Test.java:6)
finally...
```

图 5-23

可以看到结果完全一样，那么大家可能就会有疑问了，既然结果完全一样，那么为什么要使用 finally 呢？别着急，看完下面这个例子你就明白了。定义一个带有返回值的方法 test，在该方法中加入 try-catch，具体实现如代码 5-27 所示。

代码 5-27

```java
public class Test {
    public static void main(String[] args) {
        System.out.println(test());
    }
    public static int test() {
        try {
            System.out.println("try...");
```

```
        return 10;
    }catch (Exception e) {
        // TODO: handle exception
    }
    System.out.println("finally...");
    return 20;
    }
}
```

运行结果如图 5-24 所示。

```
try...
10
```

图 5-24

通过结果可以看到，try 代码块中进行了 return 操作，所以后续的代码都不会执行。现在对代码进行修改，如代码 5-28 所示。

代码 5-28

```
public class Test {
    public static void main(String[] args) {
        System.out.println(test());
    }
    public static int test() {
        try {
            System.out.println("try...");
            return 10;
        }catch (Exception e) {
            // TODO: handle exception
        }finally {
            System.out.println("finally...");
            return 20;
        }
    }
}
```

运行结果如图 5-25 所示。

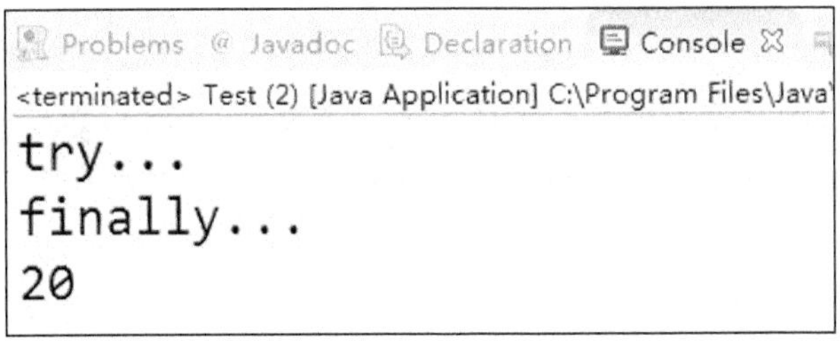

```
try...
finally...
20
```

图 5-25

通过结果可以看到，虽然 try 代码块中执行了 return 操作，但是 finally 代码块中的程序依然会执行，并且会覆盖 try 中 return 的结果，返回给外部调用者，正是因为 finally 的这个特性，一般会在 finally 中进行释放资源的操作。

5.4.3 异常类

Java 将运行时出现的错误全部封装成类，并且不是一个类，而是一组类。同时这些类之间是有层级关系的，由树状结构一层层向下分级，处在最顶端的类是 Throwable，是所有异常类的根结点。Throwable 有两个直接子类：Error 和 Exception，这两个类我们在上一节也提到过。Error 表示系统错误，程序无法解决；Exception 指程序运行时出现的错误，程序可以处理。Throwable、Error 和 Exception 都存放在 java.lang 包中。

Error 常见的子类有 VirtualMachineError、AWTError、IOError。VirtualMachineError 的常见的子类有 StackOverflowError 和 OutOfMemoryError，用来描述内存溢出等系统问题。VirtualMachineError、StackOverflowError 和 OutOfMemoryError 都存放在 java.lang 包中，AWTError 存放在 java.awt 包中，IOError 存放在 java.io 包中。

Exception 常见的子类主要有 IOException 和 RuntimeException，IOException 存放在 java.io 包中，RuntimeException 存放在 java.lang 包中。Exception 类要重点关注，因为这部分异常是需要我们在编写代码的过程中手动进行处理的。

IOException 的常用子类有 FileLockInterruptionException、FileNotFoundException 和 FilerException，这些异常通常都是处理通过 IO 流进行文件传输时发生的错误，在后面 IO 流的章节我们会详细讲解。FileLockInterruptionException 存放在 java.nio.channels 包中，FileNotFoundException 存放在 java.io 包中，FilerException 存放在 javax.annotation.processing 包中。以下这些类全部存放在 java.lang 包中，异常类的体系结构如图 5-26 所示。

- RuntimeException 的常用子类如下。
- ArithmeticException：表示数学运算异常。
- ClassNotFoundException：表示类未定义异常。
- IllegalArgumentException：表示参数格式错误异常。
- ArrayIndexOutOfBoundsException：表示数组下标越界异常。
- NullPointerException：表示空指针异常。
- NoSuchMethodError：表示方法未定义异常。
- NumberFormatException：表示将其他数据类型转为数值类型时的不匹配异常。

以上我们列举出了实际开发中经常使用到的异常类，还有很多没有列举出来，感兴趣的读者可以自己去查找 API 文档。除了使用 Java 官方提供的异常类，我们也可以根据需求自定义异常类。

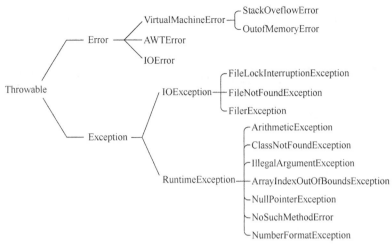

图 5-26

5.4.4 throw 和 throws

throw 和 throws 是 Java 在处理异常时使用的两个关键字，都用来抛出异常，但是使用的方式以及表示的含义完全不同，接下来就带领大家一起来学习二者的区别。Java 中抛出异常有 3 种方式，第 1 种是我们之前介绍过的使用 try-catch 代码块捕获异常。这种方式其实是一种防范机制，即代码中有可能会抛出异常。如果抛出，则捕获并进行处理；如果不抛出，则程序继续向后执行。具体实现如代码 5-29 所示。

代码 5-29

```java
public class Test {
    public static void main(String[] args) {
        String str = "Java";
        try {
            Integer num = Integer.parseInt(str);
        }catch (Exception e) {
            // TODO: handle exception
            e.printStackTrace();
        }
    }
}
```

运行结果如图 5-27 所示。

```
java.lang.NumberFormatException: For input string: "Java"
    at java.base/java.lang.NumberFormatException.forI
    at java.base/java.lang.Integer.parseInt(Unknown S
    at java.base/java.lang.Integer.parseInt(Unknown S
    at com.southwind.test.Test.main(Test.java:7)
```

图 5-27

可以看到结果抛出了类型转换的异常,不能将 String 类型的"Java"转为 int 类型的数据。同时在代码中我们不添加 try-catch 代码块,但结果是一样的,具体实现如代码 5-30 所示。

代码 5-30

```java
public class Test {
    public static void main(String[] args) {
        String str = "Java";
        Integer num = Integer.parseInt(str);
    }
}
```

运行结果如图 5-28 所示。

```
java.lang.NumberFormatException: For input string: "Java"
    at java.base/java.lang.NumberFormatException.forI
    at java.base/java.lang.Integer.parseInt(Unknown S
    at java.base/java.lang.Integer.parseInt(Unknown S
    at com.southwind.test.Test.main(Test.java:7)
```

图 5-28

可以看到结果完全一样,为什么我们不添加 try-catch 代码块,程序同样可以抛出异常呢?答案是 Java 有非常完善的错误处理机制,即使开发者不主动在程序中进行异常捕获,Java 虚拟机也会自动完成异常处理。这相当于员工因为疏忽没有把工作做好,老板会在汇总工作时发现问题并帮员工解决,保证最终的工作成果没有问题。但是很显然这种方式不是很好,不能总让老板去处理错误,员工在做本职工作时应该把所有问题都处理好。回到代码中也是一样的道理,我们在编写程序时应尽量将异常进行处理,这个工作不要交给 Java 虚拟机去处理。如果我们进行如下修改,程序就不会抛出异常,如代码 5-31 所示。

代码 5-31

```java
public class Test {
    public static void main(String[] args) {
        String str = "10";
        Integer num = Integer.parseInt(str);
    }
}
```

使用 throw 是开发者主动抛出异常,即读到 throw 代码就一定会抛出异常,基本语法:"throw new Exception();",这是一种基于代码的逻辑判断从而主动抛出异常的方式,如代码 5-32 所示。

代码 5-32

```java
public class Test {
    public static void main(String[] args) {
        String str = "Java";
```

```
        if(str.equals("Java")) {
            throw new NumberFormatException();
        }else {
            int num = Integer.parseInt(str);
        }
    }
}
```

运行结果如图 5-29 所示。

```
Exception in thread "main" java.lang.NumberFormatException
        at com.southwind.test.Test.main(Test.java:7)
```

图 5-29

在上述代码中，我们主动对 str 进行判断，如果 str 的值为"Java"，则直接抛出 Number-FormatException 异常。所以 try-catch 是捕获可能抛出的异常，throw 是确定会抛出异常，这是二者的区别。那么 throws 是如何抛出异常的呢？try-catch 和 throw 都是作用于具体的逻辑代码，throws 则是作用于方法，用来描述该方法可能会抛出的异常，具体实现如代码 5-33 所示。

代码 5-33

```java
public class Test {
    public static void main(String[] args) {
        try {
            test();
        } catch (NumberFormatException e) {
            // TODO: handle exception
            e.printStackTrace();
        }
    }
    public static void test() throws NumberFormatException{
        String str = "Java";
        int num = Integer.parseInt(str);
    }
}
```

test()方法在定义时通过 throws 关键字声明了该方法可能会抛出 NumberFormatException 异常，所以我们在调用该方法时，需要手动使用 try-catch 进行捕获。同时 catch 代码块中可以捕获 NumberFormatException，也可以捕获 Exception，这两种方式都是没问题的，如代码 5-34 所示。

代码 5-34

```java
public class Test {
    public static void main(String[] args) {
        try {
            test();
        } catch (Exception e) {
```

```
         // TODO: handle exception
         e.printStackTrace();
      }
   }
   public static void test() throws NumberFormatException{
      String str = "Java";
      int num = Integer.parseInt(str);
   }
}
```

为什么这里捕获 NumberFormatException 和 Exception 都可以呢？因为 Java 的多态特性。我们在前面的章节中介绍过面向对象的三大特征之一的多态，NumberFormatException 是具体的数值类型转换异常，Exception 是所有异常的父类。NumberFormatException 也可以理解成 Exception 的另外一种表现形式，这里我们使用这两类异常都是可以的。同时在调用 test 方法时，可以使用 try-catch 主动捕获，也可以不添加 try-catch，直接交给 Java 虚拟机来处理异常。所以这种情况下，加不加 try-catch 都是可以的，但是建议添加。既然我们在 catch 中可以使用多态来捕获异常，那么在方法定义时也可以使用多态来描述可能发生的异常，即代码可以做如下修改，如代码 5-35 所示。

代码 5-35

```
public class Test {
   public static void main(String[] args) {
      try {
         test();
      } catch (Exception e) {
         // TODO: handle exception
         e.printStackTrace();
      }
   }
   public static void test() throws Exception{
      String str = "Java";
      int num = Integer.parseInt(str);
   }
}
```

但是在这种情况下，main 方法在调用 test 方法时就必须手动进行捕获。这里需要注意，如果方法抛出 RunntimeException 异常或者其子类异常，外部调用该方法时可以不进行 try-catch 捕获。如果方法抛出的是 Exception 异常或者其子类异常，则外部调用时必须进行 try-catch 捕获，否则会报错，如图 5-30 所示。

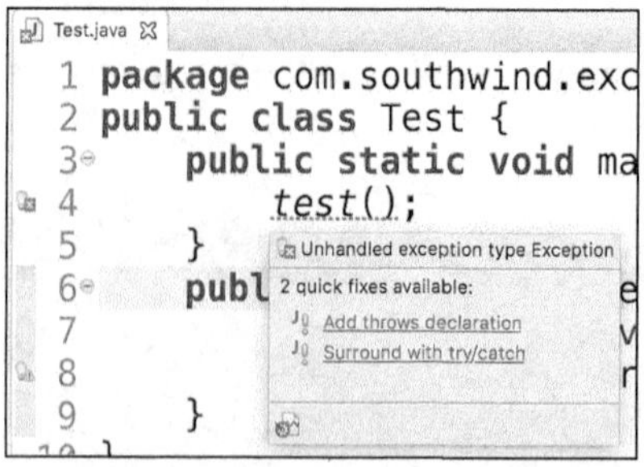

图 5-30

如果不添加 try-cath，也可以通过让 main 方法抛出该异常的方式来解决这个错误，如代码 5-36 所示。

代码 5-36

```java
public class Test {
    public static void main(String[] args) throws Exception {
        test();
    }
    public static void test() throws Exception{
        String str = "Java";
        int num = Integer.parseInt(str);
    }
}
```

test()方法声明时会抛出 Exception，主方法中的代码在调用 test()方法时就需要对异常进行处理，这里选择将异常抛出，那么抛出的异常是交给谁来处理的呢？你一定知道答案的！没错，就是交给 Java 虚拟机来处理了。

5.4.5 自定义异常类

在实际开发中，我们除了使用 Java 提供的异常类之外，也可以根据需求来自定义异常类，比如定义一个方法，对传入的参数进行++操作并返回，同时要求参数必须是整数类型，如果传入的参数不是整数类型则抛出自定义异常，具体实现如代码 5-37 所示。

代码 5-37

```java
public class MyNumberException extends Exception {
    public MyNumberException(String error) {
        super(error);
    }
}


public class Test {
    public static void main(String[] args) {
        Test test = new Test();
        try {
            int num = test.add("hello");
        } catch (MyNumberException e) {
            // TODO Auto-generated catch block
            e.printStackTrace();
        }
    }
    public int add(Object object) throws MyNumberException {
        if(!(object instanceof Integer)) {
            String error = "传入的参数不是整数类型";
            throw new MyNumberException(error);
        }else {
            int num = (int) object;
            return num++;
        }
    }
```

        }
    }

add()方法定义时声明了可能会抛出 MyNumberException 异常，是 Exception 的子类，所以在 main 方法中调用 add()方法时需要手动进行处理，上述代码我们是通过 try-catch 的方式处理的，同时也可以直接让 main 方法抛出异常，如代码 5-38 所示。

代码 5-38

```java
public class Test {
    public static void main(String[] args) throws MyNumberException {
        Test test = new Test();
        int num = test.add("hello");
    }
    public int add(Object object) throws MyNumberException {
        if(!(object instanceof Integer)) {
            String error = "传入的参数不是整数类型";
            throw new MyNumberException(error);
        }else {
            int num = (int) object;
            return num++;
        }
    }
}
```

两种方式的运行结果一样，如图 5-31 所示。

```
com.southwind.entity.MyNumberException: 传入的参数不是整数类型
        at com.southwind.test.Test.add(Test.java:19)
        at com.southwind.test.Test.main(Test.java:9)
```

图 5-31

推荐使用手动 try-catch 的方式，谁调用谁处理，不要把所有的任务都交给 Java 虚拟机来处理。这里我们需要注意，Java 中有些异常在 throw 之后，还需要在方法定义处添加 throws 声明，有些异常则不需要，直接 throw 即可。这是因为 Exception 的异常分 checked exception 和 runtime exception，checked exception 表示需要强制去处理的异常，即 throw 异常之后需要立即处理，要么自己 try-catch，要么抛给上一层去处理，否则会报错，例如 "Unhandled exception type Exception"。而 runtime exception 没有这个限制，throw 之后可以不处理。

直接继承自 Exception 的类就是 checked exception，继承自 RuntimeException 的类就是 runtime exception。我们自定义的 MyNumberExcpetion 是直接继承 Exception 的，所以需要在 add()方法定义处声明 throws MyNumberExcpetion。

5.5 综合练习

使用面向对象章节所学的知识点，重点包括封装、继承、多态、抽象、接口来完成一个汽车查询系统。

需求描述：共有 3 种类型的汽车：小轿车、大巴车、卡车，其中小轿车座位数为 4 座，大巴车座位数为 53 座，卡车座位数为 2 座，要求使用封装、继承、抽象来完成车辆的定义。

可以对车辆信息作出修改，卡车可以运货但是载重量不能超过 12 吨，使用自定义异常来处理错误，小轿车和大巴车没有此功能，要求使用接口来实现。

具体实现如代码 5-39 所示。

代码 5-39

```java
public abstract class Car {
    private String name;
    private String color;
    public String getName() {
        return name;
    }
    public String getColor() {
        return color;
    }
    public Car(String name,String color) {
        this.name = name;
        this.color = color;
    }
    public abstract String seatNum();
}


public class Sedan extends Car{
    public Sedan(String name, String color) {
        super(name, color);
    }
    @Override
    public String seatNum() {
        // TODO Auto-generated method stub
        return "4座";
    }
}


public class Bus extends Car{
    public Bus(String name, String color) {
        super(name, color);
    }
    @Override
    public String seatNum() {
        // TODO Auto-generated method stub
        return "53座";
    }
}


public interface Container {
    public int getweight();
```

```java
    }


    public class Truck extends Car implements Container {
        private int weight;
        public Truck(String name, String color,int weight) {
            super(name, color);
            this.weight = weight;
        }
        public int getweight() {
            // TODO Auto-generated method stub
            return this.weight;
        }
        @Override
        public String seatNum() {
            // TODO Auto-generated method stub
            return "2 座";
        }
    }


    public class CarException extends Exception{
        public CarException(String message){
            super(message);
        }
    }


    public class CarTest {
        private static Scanner scanner;
        private static Sedan sedan;
        private static Bus bus;
        private static Truck truck;
        private static Car[] cars;
        static {
            scanner = new Scanner(System.in);
            sedan = new Sedan("小轿车","黑色");
            bus = new Bus("大巴车","绿色");
            truck = new Truck("卡车","蓝色",2);
            cars = new Car[3];
            cars[0] = sedan;
            cars[1] = bus;
            cars[2] = truck;
        }
        public void showCars(){
            System.out.println("车辆名称\t\t 车辆颜色\t\t 座位数\t\t 载重量");
            for(Car car : cars){
                if(car instanceof Sedan){
                    Sedan sedan = (Sedan)car;
            System.out.println(sedan.getName()+"\t\t"+sedan.getColor()+"\t\t"+sedan.seatNum());
                }
                if(car instanceof Bus){
                    Bus bus = (Bus)car;
            System.out.println(bus.getName()+"\t\t"+bus.getColor()+"\t\t"+bus.seatNum());
                }
```

```java
            if(car instanceof Truck){
                Truck truck = (Truck)car;
                System.out.println(truck.getName()+"\t\t"+truck.getColor()+"\t\t"+truck.seatNum()+"\t\t"+truck.getweight()+"吨");
            }
        }
        System.out.println("1.小轿车\t2.大巴车\t3.卡车");
        System.out.print("请选择要修改的车辆：");
        int num = scanner.nextInt();
        switch(num){
        case 1:
            update("sedan");
            break;
        case 2:
            update("bus");
            break;
        case 3:
            update("truck");
            break;
        default:
            System.out.println("车辆不存在！");
        }
    }
    public void update(String type){
        String name = null;
        String color = null;
        if(type.equals("sedan")){
            System.out.print("请输入车辆的名称：");
            name = scanner.next();
            System.out.print("请输入车辆的颜色：");
            color = scanner.next();
            Sedan sedan = new Sedan(name,color);
            cars[0] = sedan;
        }
        if(type.equals("bus")){
            System.out.print("请输入车辆的名称：");
            name = scanner.next();
            System.out.print("请输入车辆的颜色：");
            color = scanner.next();
            Bus bus = new Bus(name,color);
            cars[1] = bus;
        }
        if(type.equals("truck")){
            System.out.print("请输入车辆的名称：");
            name = scanner.next();
            System.out.print("请输入车辆的颜色：");
            color = scanner.next();
            System.out.print("请输入载重量：");
            int weight = scanner.nextInt();
            if(weight > 12){
                try {
                    throw new CarException("卡车的载重量不能超过12吨");
                } catch (CarException e) {
                    // TODO Auto-generated catch block
                    e.printStackTrace();
                    return;
                }
```

```
            }
            Truck truck = new Truck(name, color, weight);
            cars[2] = truck;
        }
        showCars();
    }
    public static void main(String[] args) {
        CarTest carTest = new CarTest();
        carTest.showCars();
    }
}
```

运行结果如图 5-32～图 5-35 所示。

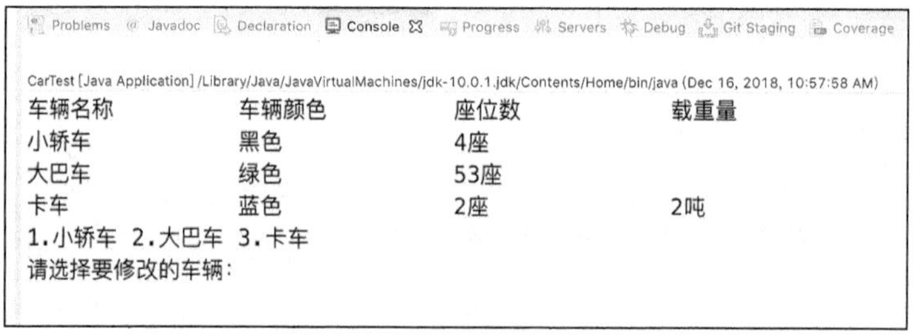

图 5-32

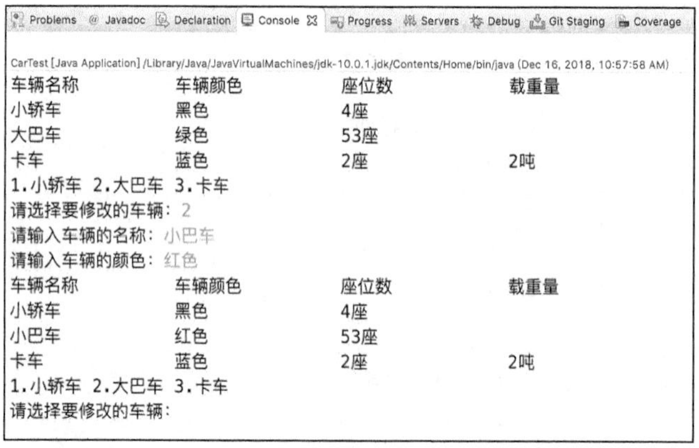

图 5-33

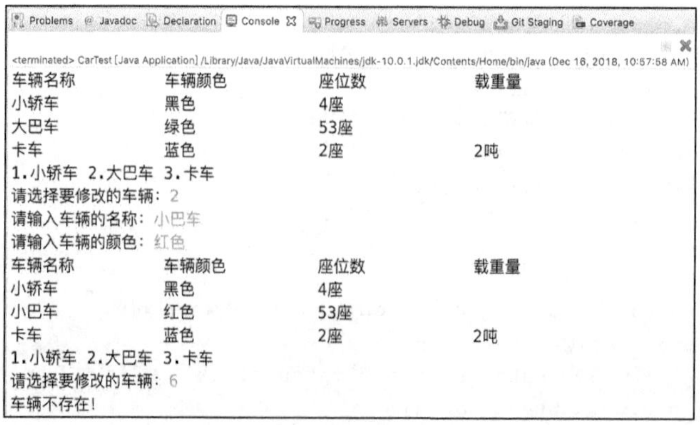

图 5-34

```
Problems  Javadoc  Declaration  Console ☒  Progress  Servers  Debug  Git Staging  Coverage
<terminated> CarTest [Java Application] /Library/Java/JavaVirtualMachines/jdk-10.0.1.jdk/Contents/Home/bin/java (Dec 16, 2018, 11:00:26 AM)
车辆名称          车辆颜色          座位数          载重量
小轿车            黑色              4座
大巴车            绿色              53座
卡车              蓝色              2座             2吨
1.小轿车 2.大巴车 3.卡车
请选择要修改的车辆: 3
请输入车辆的名称: 大卡车
请输入车辆的颜色: 红色
请输入载重量: 16
com.southwind.test.MyException: 卡车的载重量不能超过12吨
        at com.southwind.test.CarTest.update(CarTest.java:89)
        at com.southwind.test.CarTest.showCars(CarTest.java:54)
        at com.southwind.test.CarTest.main(CarTest.java:24)
```

图 5-35

5.6 小结

本章继续为大家讲解了面向对象的高级部分，包括 Object 类、包装类、接口和异常。其中 Object 类是所有 Java 类的父类，定义了 Java 体系的基础资料，通过继承传递给 Java 的每一个类，通过方法重写和多态让整个 Java 体系具有很强的灵活性。包装类是 Java 为基本数据类型提供封装的一组类，通过包装类我们可以将基本数据类型转为对象，这一点在面向对象编程中很关键。接口是抽象类的扩展，是 Java 中实现多态的重要方式，可以降低程序的耦合性，让程序变得更加灵活多变。接口就相当于零件，我们可以自由地将这些零件进行组装、整合。异常是 Java 中处理错误的一种机制，同样是基于面向对象的思想，将错误抽象成对象然后进行处理，这里需要注意的是对异常相关的几个关键字的使用。

第3部分 Java高级应用

第 6 章 多线程

> 多线程是提升程序性能非常重要的一种方式,也是我们必须要掌握的技术。使用多线程可以让程序充分利用 CPU 的资源,提高 CPU 的使用效率,从而解决高并发带来的负载均衡问题。它的优点是显而易见的:(1)资源得到更合理的利用;(2)程序设计更加简洁;(3)程序响应更快,运行效率更高。
>
> 任何一门技术都没有绝对的好与坏,有其优点就必有其缺点,多线程也存在一些缺点:(1)需要更多的内存空间来支持多线程;(2)多线程并行访问的情况可能会影响数据的准确性;(3)数据被多线程共享,可能会出现死锁的情况。
>
> 我们在实际开发的过程中,应该将程序设计得更加合理有效,避免多线程的缺点,将其优点发挥出来,从而提高程序的性能。

6.1 进程与线程

说到线程,就必须先提一个概念:进程。什么是进程?简单来理解,进程就是计算机正在运行的一个独立的应用程序,例如打开 Eclipse 编写 Java 程序就是一个进程,打开浏览器查找学习资料就是一个进程等。一个应用程序至少有一个进程。那什么是线程呢?进程和线程之间的关系是什么?线程是组成进程的基本单位,可以完成特定的功能,一个进程是由一个或多个线程组成的。

进程和线程是应用程序在执行过程中的概念,如果应用程序没有执行,比如 Eclipse 工具没有运行起来,那么就不存在进程和线程的概念。应用程序是静态的概念,进程和线程是动态概念,有创建就有销毁,存在也是暂时的,不是永久性的。进程与线程的区别在于进程在运行时拥有独立的内存空间,即每个进程所占用的内存都是独立的,互不干扰。而多个线程是共享内存空间的,但是每个线程的执行是相互独立的,同时线程必须依赖于进程才能执行,单独的线程是无法执行的,由进程来控制多个线程的执行。

了解完进程与线程的区别,接下来说说什么是多线程。我们通常所说的多线程是指在一个进程中,多个线程同时执行。这里说的同时执行不是真正意义上的同时执行,系统会自动为每个线程分配 CPU 资源，在某个具体时间段内 CPU 的资源被一个线程占用，在不同的时间段内由不同的线程来占用 CPU 资源，所以多个线程还是在交替执行，只不过因为 CPU 运行速度太快，感觉上是在同时执行。我们之前写的 Java 程序都是单线程的，Java 程序运行起来就是一个进程，该进程中只有一个线程在运行，如代码 6-1 所示。

代码 6-1

```java
public class Test {
   public static void main(String[] args) {
      System.out.println("Thread");
   }
}
```

程序运行起来之后只有一个线程在执行，就是 main 方法，所以 main 方法也叫作程序的主线程，main 方法中无论调用多少个其他类的方法，也都只是一个线程，如代码 6-2 所示。

代码 6-2

```java
public class MyTest {
   public void test() {
      for (int i = 0; i < 10; i++) {
         System.out.println("--------------MyTest");
      }
   }
}


public class Test {
   public static void main(String[] args) {
      for(int i = 0; i < 10; i++) {
         System.out.println("++++++++++++++Test");
      }
      MyTest myTest = new MyTest();
      myTest.test();
   }
}
```

运行结果如图 6-1 所示。

通过结果可以看到，我们在 main 方法中调用了两个循环逻辑，这两个循环逻辑是顺序执行的，先执行完打印"++++++++++++++Test"的循环，再执行打印"--------------MyTest"的循环。其实还是一个线程，那什么是多线程呢？两个循环逻辑不是顺序执行，而是同时执行，如图 6-2 所示。

你可以看到两个打印语句在交替输出，说明两个循环逻辑是同时在执行的，这种情况就是两个线程在同时运行。我们如何来判断程序是单线程还是多线程呢？只需要分析程序有几条分支即可，如图 6-3 和图 6-4 所示。

6.1 进程与线程

```
Problems  @ Javadoc  Declaration  Console  Progress  Servers
<terminated> Test (2) [Java Application] C:\Program Files\Java\jre-10.0.2\bin\javaw.exe
+++++++++++++++Test
+++++++++++++++Test
+++++++++++++++Test
+++++++++++++++Test
+++++++++++++++Test
+++++++++++++++Test
+++++++++++++++Test
+++++++++++++++Test
+++++++++++++++Test
+++++++++++++++Test
---------------MyTest
---------------MyTest
---------------MyTest
---------------MyTest
---------------MyTest
---------------MyTest
---------------MyTest
---------------MyTest
---------------MyTest
---------------MyTest
```

图 6-1

```
Problems  @ Javadoc  Declaration  Console  Progress  Servers
<terminated> Test (3) [Java Application] C:\Program Files\Java\jre-10.0.2\bin\javaw.exe
+++++++++++++++Test
+++++++++++++++Test
---------------MyTest
+++++++++++++++Test
---------------MyTest
---------------MyTest
---------------MyTest
---------------MyTest
+++++++++++++++Test
+++++++++++++++Test
---------------MyTest
+++++++++++++++Test
+++++++++++++++Test
+++++++++++++++Test
```

图 6-2

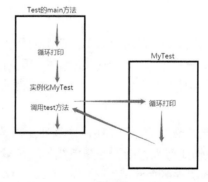

图 6-3

整个程序的执行是一条回路，所以程序只有一个线程。

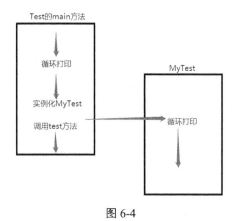

图 6-4

程序有两条回路，同时向下运行，这种情况就是多线程，两个线程同时执行。

6.2 Java 中线程的使用

6.2.1 继承 Thread 类

Java 中实现多线程的常用方式有两种：继承 Thread 类和实现 Runnable 接口，继承 Thread 类的实现分为两步：（1）创建自定义类并继承 Thread；（2）重写 Thread 的 run() 方法，并编写该线程的业务逻辑代码。具体实现如代码 6-3 所示。

代码 6-3

```
public class MyThread extends Thread{
   @Override
   public void run() {
      // TODO Auto-generated method stub
      for (int i = 0; i < 10; i++) {
         System.out.println("--------------MyThread");
      }
   }
}
```

线程创建好之后，如何调用呢？具体实现如代码 6-4 所示。

代码 6-4

```
public class Test {
   public static void main(String[] args) {
      MyThread myThread = new MyThread();
      myThread.start();
      for(int i = 0; i < 10; i++) {
         System.out.println("+++++++++++++Test");
      }
```

        }
    }

运行结果如图 6-5 所示。

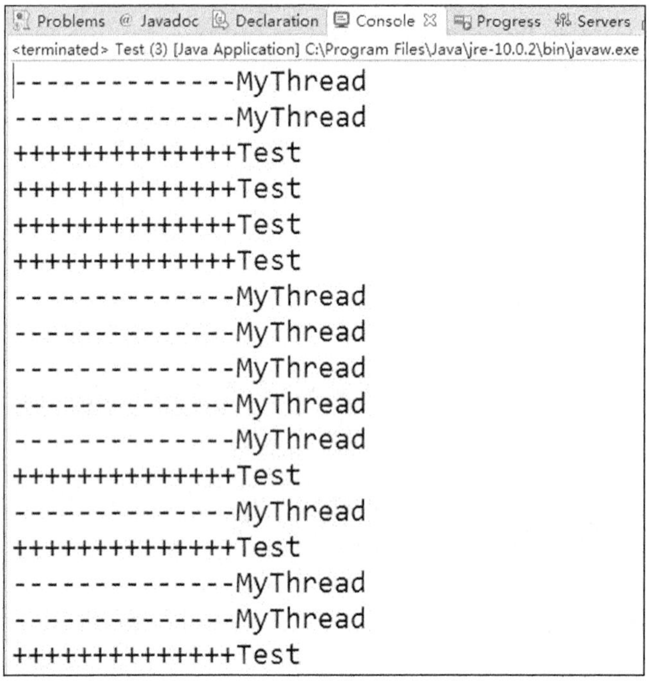

```
---------------MyThread
---------------MyThread
++++++++++++++Test
++++++++++++++Test
++++++++++++++Test
++++++++++++++Test
---------------MyThread
---------------MyThread
---------------MyThread
---------------MyThread
---------------MyThread
++++++++++++++Test
---------------MyThread
++++++++++++++Test
---------------MyThread
---------------MyThread
++++++++++++++Test
```

图 6-5

你可以看到两个循环在交替执行，当前程序中有两个线程，一个是主线程，即 main 方法，主线程完成了两件事，第一件事是开启了一个子线程 MyThread，第二件事是执行一个循环操作，因为子线程开启之后和主线程在竞争 CPU 资源，交替执行，所以会看到交替打印两个循环信息的结果。这里需要注意，开启子线程是通过调用线程对象的 start() 方法来完成的，一定不能调用 run() 方法，调用 run() 方法是普通的方法调用，相当于在主线程中顺序执行了两个循环，并没有开启一个可以和主线程抢占 CPU 资源的子线程。所以调用 run() 方法并不是多线程，还是一个主线程在执行，如代码 6-5 所示。

代码 6-5

```
public class Test {
    public static void main(String[] args) {
        MyThread myThread = new MyThread();
        myThread.run();
        for(int i = 0; i < 10; i++) {
            System.out.println("++++++++++++++Test");
        }
    }
}
```

运行结果如图 6-6 所示。

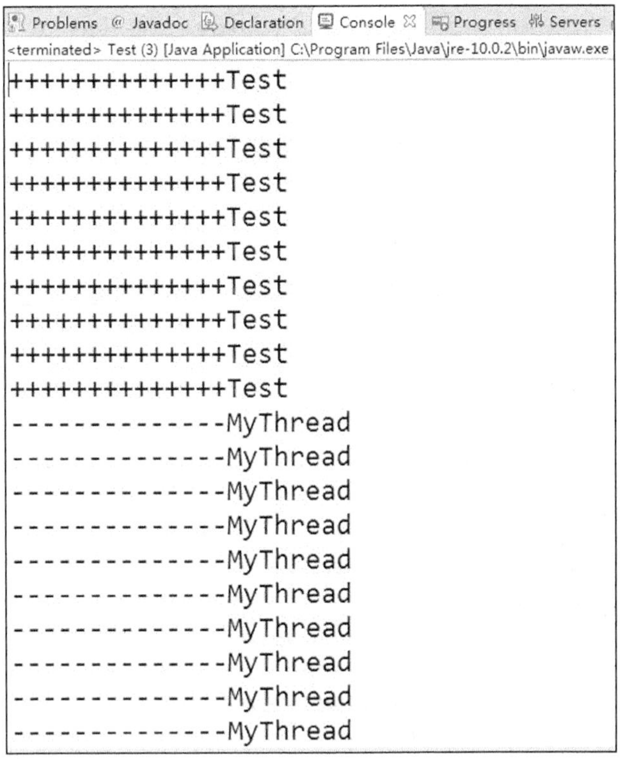
Problems @ Javadoc Declaration Console Progress Servers
<terminated> Test (3) [Java Application] C:\Program Files\Java\jre-10.0.2\bin\javaw.exe
++++++++++++++Test
++++++++++++++Test
++++++++++++++Test
++++++++++++++Test
++++++++++++++Test
++++++++++++++Test
++++++++++++++Test
++++++++++++++Test
++++++++++++++Test
++++++++++++++Test
--------------MyThread
--------------MyThread
--------------MyThread
--------------MyThread
--------------MyThread
--------------MyThread
--------------MyThread
--------------MyThread
--------------MyThread
--------------MyThread

图 6-6

通过结果可以看到，并不是多线程在并行执行。

6.2.2 实现 Runnable 接口

Java 中创建多线程的另外一种方式是通过 Runnable 接口的实现来完成，具体分为两步：（1）创建自定义类并实现 Runnable 接口；（2）实现 run()方法，编写该线程的业务逻辑代码。具体实现如代码 6-6 所示。

代码 6-6

```
public class MyRunnable implements Runnable{
   @Override
   public void run() {
      // TODO Auto-generated method stub
      for (int i = 0; i < 10; i++) {
         System.out.println("--------------MyRunnable");
      }
   }
}
```

调用 MyRunnable 线程的具体实现如代码 6-7 所示。

代码 6-7

```
public class Test {
   public static void main(String[] args) {
```

```java
        MyRunnable myRunnable = new MyRunnable();
        Thread thread = new Thread(myRunnable);
        thread.start();
        for(int i = 0; i < 10; i++) {
            System.out.println("++++++++++++++Test");
        }
    }
}
```

MyRunnable 的使用与 MyThread 略有不同，MyRunnable 相当于定义了线程业务逻辑，它本身并不是线程对象，所以还需要实例化 Thread 对象，然后将 MyRunnable 对象赋给 Thread 对象。这样 Thread 对象就知道了它要完成的业务逻辑，再通过调用 Thread 对象的 start()方法来启动该线程，运行结果如图 6-7 所示。

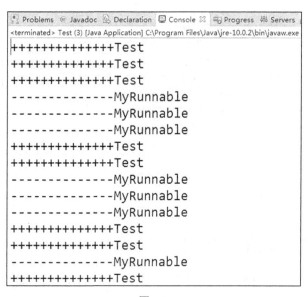

图 6-7

6.2.3 线程的状态

线程共有 5 种状态，在特定的情况下，线程可以在不同的状态之间切换，5 种状态如下所示。

- 创建状态：实例化了一个新的线程对象，还未启动。
- 就绪状态：创建好的线程对象调用 start()方法完成启动，进入线程池等待抢占 CPU 资源。
- 运行状态：线程对象获取了 CPU 资源，在一定的时间内执行任务。
- 阻塞状态：正在运行的线程暂停执行任务，释放所占用的 CPU 资源。并在解除阻塞之后也不能直接回到运行状态，而是重新回到就绪状态，等待获取 CPU 资源。
- 终止状态：线程运行完毕或因为异常导致该线程终止运行。

线程状态之间的转换如图 6-8 所示。

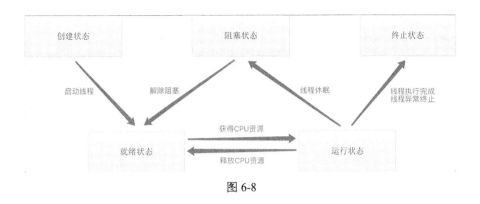

图 6-8

6.3 线程调度

6.3.1 线程休眠

休眠指让当前线程暂停执行，从运行状态进入阻塞状态，将 CPU 资源让给其他线程的一种调度方式，要通过调用 sleep() 方法来实现。sleep(long millis) 是 java.lang.Thread 类中定义的方法，使用时需要指定当前线程休眠的时间，传入一个 long 类型的数据作为休眠时间，单位为毫秒。任意一个线程的实例化对象都可以调用该方法，如代码 6-8 所示。

代码 6-8

```
public class MyThread extends Thread{
   @Override
   public void run() {
      // TODO Auto-generated method stub
      for (int i = 0; i < 10; i++) {
         if(i == 5) {
            try {
               sleep(1000);
            } catch (InterruptedException e) {
               // TODO Auto-generated catch block
               e.printStackTrace();
            }
         }
         System.out.println("--------------MyThread");
      }
   }
}
```

代码 6-8 表示当循环执行到第 6 次即 i=5 时，会休眠 1000 毫秒，1000 毫秒之后线程进入就绪状态，重新等待系统为其分配 CPU 资源。这是在线程内部执行休眠操作，也可以在外部使用线程时执行休眠操作，如代码 6-9 所示。

代码 6-9

```
public class MyThread extends Thread{
```

```java
    @Override
    public void run() {
        // TODO Auto-generated method stub
        for (int i = 0; i < 10; i++) {
            System.out.println("--------------MyThread");
        }
    }
}

public class Test {
    public static void main(String[] args) {
        MyThread myThread = new MyThread();
        try {
            myThread.sleep(1000);
        } catch (InterruptedException e) {
            // TODO Auto-generated catch block
            e.printStackTrace();
        }
        mt.start();
        for(int i = 0; i < 100; i++) {
            System.out.println("+++++++++++++Test");
        }
    }
}
```

代码 6-9 表示在 Test 类的主线程中创建 MyThread 子线程，程序运行时 MyThread 子线程休眠 1000 毫秒之后再启动。

前面我们说到任意一个线程的实例化对象都可以调用 sleep()方法，即每一个线程都可以进行休眠。那么如何让主线程休眠呢？主线程并不是一个我们手动实例化的线程对象，不能直接调用 sleep()方法，这种情况下可以通过 Threa 类的静态方法 currentThread 来获取主线程对应的线程对象，然后调用 sleep()方法，如代码 6-10 所示。

代码 6-10

```java
public class Test {
    public static void main(String[] args) {
        for(int i = 0; i < 10; i++) {
            if(i == 5) {
                try {
                    Thread.currentThread().sleep(1000);
                } catch (InterruptedException e) {
                    // TODO Auto-generated catch block
                    e.printStackTrace();
                }
            }
            System.out.println("+++++++++++++Test");
        }
    }
}
```

代码 6-10 表示主线程循环执行到第 6 次即 i=5 时，会进行休眠状态，然后暂停执行，等待 1000 毫秒之后进入就绪状态，并等待获取 CPU 资源从而进入运行状态继续执行。无论通过哪种方式调用 sleep() 方法都需要注意处理异常，因为 sleep() 方法在定义时声明了可能会抛出的异常 InterruptedException，如代码 6-11 所示。

代码 6-11

```
public static native void sleep(long millis) throws InterruptedException;
```

所以在外部调用 sleep() 方法时就必须处理可能抛出的异常，这里给出两种方案：（1）通过 try-catch 主动捕获；（2）main 方法定义处抛出该异常交给 JVM 去处理。推荐使用第 1 种方案。

6.3.2 线程合并

合并的意思是将指定的某个线程加入到当前线程中，合并为一个线程，由两个线程交替执行变成一个线程中的两个子线程顺序执行，即一个线程执行完毕之后再来执行第二个线程，通过调用线程对象的 join() 方法来实现合并。具体是如何来合并的呢，谁为主谁为从？假设有两个线程：线程甲和线程乙。线程甲在执行到某个时间点的时候调用线程乙的 join() 方法，则表示从当前时间点开始 CPU 资源被线程乙独占，线程甲进入阻塞状态。直到线程乙执行完毕，线程甲进入就绪状态，等待获取 CPU 资源进入运行状态继续执行，如代码 6-12 所示。

代码 6-12

```java
public class JoinRunnable implements Runnable{
    @Override
    public void run() {
        // TODO Auto-generated method stub
        for(int i = 0 ; i < 20; i++) {
            System.out.println(i+"----------JoinRunnable");
        }
    }
}


public class Test {
    public static void main(String[] args) {
        JoinRunnable joinRunnable = new JoinRunnable();
        Thread thread = new Thread(joinRunnable);
        thread.start();
        for(int i = 0; i < 100; i++) {
            if(i == 10) {
                try {
                    thread.join();
                } catch (InterruptedException e) {
                    // TODO Auto-generated catch block
                    e.printStackTrace();
                }
            }
            System.out.println(i+"+++++++++main");
        }
    }
}
```

通过实现接口的方式定义 JoinRunnable，在 Test 类的主线程中开启子线程。当主线程循环执行到 i==10 的节点时，将子线程合并到主线程中，此时主线程进入阻塞状态。而后子线程执行，当子线程执行完毕之后，主线程继续执行，运行结果如图 6-9 所示。

```
7++++++++++main
7----------JoinRunnable
8++++++++++main
8----------JoinRunnable
9++++++++++main
9----------JoinRunnable
10----------JoinRunnable
11----------JoinRunnable
12----------JoinRunnable
13----------JoinRunnable
14----------JoinRunnable
15----------JoinRunnable
16----------JoinRunnable
17----------JoinRunnable
18----------JoinRunnable
19----------JoinRunnable
10++++++++++main
11++++++++++main
12++++++++++main
```

图 6-9

join()方法存在重载：join(long millis)，如果某个时间点在线程甲中调用了线程乙的 sleep(1000)方法，表示从当前这一时刻起，线程乙会独占 CPU 资源，线程甲进入阻塞状态。当线程乙执行了 1000 毫秒之后，线程甲重新进入就绪状态。

同样是完成线程合并的操作，join()和 join(long millis)还是有区别的，join()表示在被调用线程执行完成之后才能执行其他线程。join(long millis)则表示被调用线程执行 millis 毫秒之后，无论是否执行完毕，其他线程都可以和它来争夺 CPU 资源。join(long millis)的具体使用如代码 6-13 所示。

代码 6-13

```java
public class JoinRunnable implements Runnable{
    @Override
    public void run() {
        // TODO Auto-generated method stub
        for(int i = 0 ; i < 20; i++) {
            try {
                Thread.currentThread().sleep(1000);
            } catch (InterruptedException e) {
                // TODO Auto-generated catch block
                e.printStackTrace();
            }
            System.out.println(i+"----------JoinRunnable");
        }
```

```java
        }
    }

public class Test {
    public static void main(String[] args) {
        JoinRunnable joinRunnable = new JoinRunnable();
        Thread thread = new Thread(joinRunnable);
        thread.start();
        for(int i = 0; i < 100; i++) {
            if(i == 10) {
                try {
                    thread.join(3000);
                } catch (InterruptedException e) {
                    // TODO Auto-generated catch block
                    e.printStackTrace();
                }
            }
            System.out.println(i+"++++++++++main");
        }
    }
}
```

运行结果如图 6-10 所示。

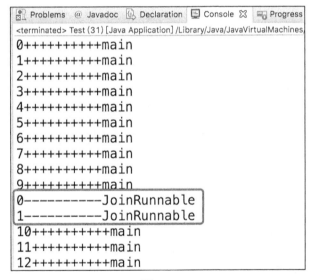

图 6-10

通过结果可以看到当主线程执行到 i==10 时,子线程会合并到主线程中,并且是通过调用 join(3000)方法完成合并的。所以从此刻开始子线程独占 CPU 资源执行 3000 毫秒之后,主线程继续与其抢占资源。因为子线程每次执行都会休眠 1000 毫秒,所以看到的结果是执行了两次子线程之后,主线程再次进入就绪状态来抢占 CPU 资源。

6.3.3 线程礼让

线程礼让是指在某个特定的时间点,让线程暂停抢占 CPU 资源的行为,即从运行状态或就绪状态来到阻塞状态,从而将 CPU 资源让给其他线程来使用。这相当于现实生活中地铁排队进站,排到该你进站的时候,你让其他人先进,把这次进站的机会让给其他人。但是这并不意味着你放弃排队,你只是在某个时间点做了一次礼让,过了这个时间点,你依然要参与到排队的序列中。线程中的礼让也是如此,假如有线程甲和线程乙在交替执行,在某个时间点线程甲做出了礼让,所以在这个时间节点线程乙拥有了 CPU 资源,执行其业务逻辑,但不是说线程甲会一直暂停执行,直到线程乙执行完毕再来执行线程甲。线程甲只是在特定的时间节点礼让,过了这个时间节点,线程甲再次进入就绪状态,和线程乙争夺 CPU 资源。Java 中的线程礼让,通过调用 yield()方法完成,具体实现如代码 6-14 所示。

代码 6-14

```java
public class YieldThread1 extends Thread{
    @Override
    public void run() {
        // TODO Auto-generated method stub
        for (int i = 0; i < 10; i++) {
            if(i == 5) {
                Thread.currentThread().yield();
            }
            System.out.println(Thread.currentThread().getName()+"------"+i);
        }
    }
}


public class YieldThread2 extends Thread{
    @Override
    public void run() {
        // TODO Auto-generated method stub
        for (int i = 0; i < 10; i++) {
            System.out.println(Thread.currentThread().getName()+"------"+i);
        }
    }
}


public class Test {
    public static void main(String[] args) {
        YieldThread1 thread1 = new YieldThread1();
        thread1.setName("Thread-1");
        YieldThread2 thread2 = new YieldThread2();
        thread2.setName("Thread-2");
        thread1.start();
        thread2.start();
    }
}
```

定义两个线程类 YieldThread1 和 YieldThread2,循环执行输出语句,并且 YieldThread1 中的循环执行到 i=5 时,会做出礼让,将 CPU 资源让给 YieldThread2。通过调用 setName()方法可以给线程对象自定义名称,在测试类中创建两个线程对象并启动,运行结果如图 6-11 所示。

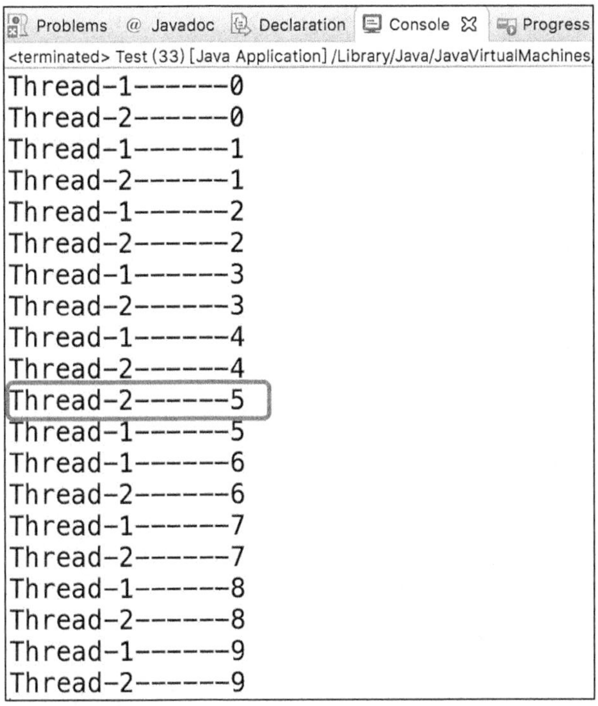

```
Problems  @ Javadoc  Declaration  Console ⊠  Progress
<terminated> Test (33) [Java Application] /Library/Java/JavaVirtualMachines
Thread-1------0
Thread-2------0
Thread-1------1
Thread-2------1
Thread-1------2
Thread-2------2
Thread-1------3
Thread-2------3
Thread-1------4
Thread-2------4
Thread-2------5
Thread-1------5
Thread-1------6
Thread-2------6
Thread-1------7
Thread-2------7
Thread-1------8
Thread-2------8
Thread-1------9
Thread-2------9
```

图 6-11

可以看到，前半段的 YieldThread1 和 YieldThread2 是在交替执行。当 YieldThread1 的 i=5 时，它暂停了执行，YieldThread2 开始执行，但是从下一个时刻起，YieldThread1 又参与到 CPU 资源的争夺中。

6.3.4 线程中断

前面几个章节我们重点介绍了线程调度的方式，本节我们来学习线程中断的实现。有多种情况可以造成线程停止运行，例如线程执行完毕之后会自动停止该线程。线程执行过程中遇到错误会抛出异常并停止该线程，线程在执行过程中会根据需求手动停止该线程。我们要介绍的线程中断就是第 3 种情况,线程在执行过程中,通过手动操作来停止该线程。比如当用户在执行一个操作时，因为网络问题导致延迟，则对应的线程对象就一直处于运行状态，如果用户希望结束这个操作，即终止该线程，此时我们就要使用线程中断机制了。

Java 中实现线程中断机制有如下几个常用方法：

- public void stop()
- public void interrupt()
- public boolean isInterrupted()

其中 stop 方法在新版本的 JDK 中已经不推荐使用了，所以我们这里不再对 stop 方法进行讲解，重点关注其他两个方法。interrupt 是一个实例方法，当一个线程对象调用该方法时，表示中断当前线程对象。每个线程对象都是通过一个标志位来判断当前是否为中断状态，isInterrupted()方法就是用来获取当前线程对象的标志位的。true 表示清除了标志位，当前线程对象已经中断；false 表示没有清除标志位，当前对象没有中断。当一个线程对象处于不同的状态时，中断机制也是不同的，接下来我们分别演示不同状态下的线程中断。

创建状态：实例化线程对象，但并未启动，如代码 6-15 所示。

代码 6-15

```java
public class Test {
    public static void main(String[] args) {
        Thread thread = new Thread();
        //获取当前线程对象的状态
        System.out.println(thread.getState());
        thread.interrupt();
        System.out.println(thread.isInterrupted());
    }
}
```

getState()方法可以获取当前线程对象的状态，实例化一个线程对象 thread，但是并未启动该对象，直接中断，运行结果如图 6-12 所示。

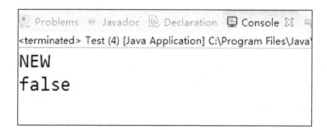

图 6-12

可以看到结果，NEW 表示当前线程对象为创建状态，flase 表示的是当前线程并未中断。因为当前线程状态根本就没有启动，所以就不存在中断，不需要清除标志位，所以 isInterrupted 的返回值为 false。

运行状态：实例化线程对象，启动该线程，循环输出语句，当 i=5 时中断线程，如代码 6-16 所示。

代码 6-16

```java
public class Test {
    public static void main(String[] args) {
        Thread thread = new Thread(new Runnable() {
            @Override
            public void run() {
                // TODO Auto-generated method stub
                for (int i = 0; i < 10; i++) {
                    if(i == 5) {
                        Thread.currentThread().interrupt();
                    }
                    System.out.println("++++++++++Test");
                }
            }
```

```
        });
        thread.start();
        System.out.println(thread.getState());
        System.out.println(thread.isInterrupted());
        System.out.println(thread.getState());
    }
}
```

运行结果如图 6-13 所示。

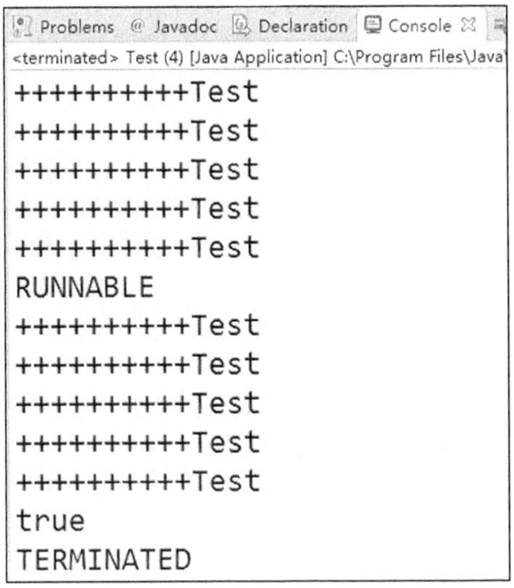

Problems
Javadoc
Declaration
Console
<terminated> Test (4) [Java Application] C:\Program Files\Java
++++++++++Test
++++++++++Test
++++++++++Test
++++++++++Test
++++++++++Test
RUNNABLE
++++++++++Test
++++++++++Test
++++++++++Test
++++++++++Test
++++++++++Test
true
TERMINATED

图 6-13

6.4 线程同步

6.4.1 线程同步的实现

Java 中允许多线程并行访问，即同一个时间段内多个线程同时完成各自的操作。这样就会带来一个问题，当多个变量同时操作一个共享数据时，可能会导致数据不准确的问题。例如，我们统计多线程用户访问量，并输出访问信息，如代码 6-17 所示。

代码 6-17

```
public class Account implements Runnable{
    private static int num;
    @Override
    public void run() {
        // TODO Auto-generated method stub
        try {
            Thread.currentThread().sleep(1);
        } catch (InterruptedException e) {
            // TODO Auto-generated catch block
            e.printStackTrace();
        }
```

```
            num++;
            System.out.println(Thread.currentThread().getName()+"是当前的第"+num+"位访客");
        }
    }

    public class AccountTest {
        public static void main(String[] args) {
            Account account = new Account();
            Thread t1 = new Thread(account,"线程1");
            Thread t2 = new Thread(account,"线程2");
            t1.start();
            t2.start();
        }
    }
```

运行结果如图 6-14 所示。

图 6-14

可以看到此时的访问量数据是有问题的，线程 1 和线程 2 都显示为第 2 位访客，什么原因导致数据出错的呢？就是因为两个线程在同时访问静态资源 num，如图 6-15 所示。

图 6-15

这就是多线程同时访问共享数据时带来的隐患，项目中一定要解决这个问题。比如金融项目，涉及资金安全的对这部分要求是非常严格的。那么如何来解决这个问题呢？需要使用线程同步。可以通过 synchronized 修饰方法来实现线程同步，每个 Java 对象都有一个内置锁，内置锁会保护使用 synchronized 关键字修饰的方法，要调用该方法必须先获得内置锁，否则就处于阻塞状态，具体实现如代码 6-18 所示。

代码 6-18

```
public class Account implements Runnable{
```

```
    private static int num;
    @Override
    public synchronized void run() {
        // TODO Auto-generated method stub
        try {
            Thread.currentThread().sleep(1);
        } catch (InterruptedException e) {
            // TODO Auto-generated catch block
            e.printStackTrace();
        }
        num++;
        System.out.println(Thread.currentThread().getName()+"是当前的第"+num+"位访客");
    }
}
```

再次运行程序，结果如图 6-16 所示。

图 6-16

通过结果可以看到现在的访问量数据是正确的，就是因为实现了线程同步的机制，如图 6-17 所示。

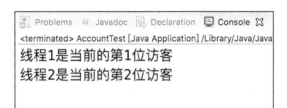

图 6-17

假设线程 1 先到，获取了 run()方法的锁，之后线程 2 到了，发现 run()方法被锁起来了。要调用方法必须拿到锁，但是此时锁被线程 1 获取了，只有当线程 1 调用完方法之后才会释放锁。执行了一次 num++，然后输出信息，之后线程 1 释放内置锁。线程 2 获取到了内部锁，由阻塞状态进入运行状态，调用 run()方法，再一次执行 num++并输出信息。两个线程是按照先后顺序来执行的，并没有同时去修改 num，所以看到了正确的结果。

synchronized 可以修饰实例方法，也可以修饰静态方法，但是两者在使用上是有区别的。接下来，我们通过实际的例子来学习，现有程序如代码 6-19 所示。

代码 6-19

```java
public class SynchronizedTest {
    public static void main(String[] args) {
        for(int i = 0; i < 5;i++) {
            Thread thread = new Thread(new Runnable() {
                @Override
                public void run() {
                    // TODO Auto-generated method stub
                    SynchronizedTest.test();
                }
            });
            thread.start();
        }
    }
    public static void test() {
        System.out.println("start...");
        try {
            Thread.currentThread().sleep(1000);
        } catch (InterruptedException e) {
            // TODO Auto-generated catch block
            e.printStackTrace();
        }
        System.out.println("end...");
    }
}
```

运行程序，结果如图 6-18 所示。

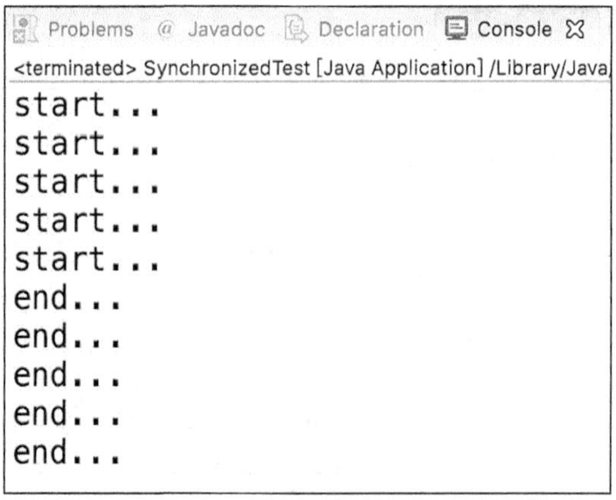

图 6-18

程序运行结果并不是"start..." "end..."成对输出，而是先输出所有的"start..."再输出所有的"end..."。这是什么原因造成的？就是因为多线程并行访问，test()方法中输出"start..."之后，休眠了 1000ms，所以就给了其他线程执行的机会。1000ms 执行 5 次线程时间上绰绰有余，所以实际的运行情况是 5 个线程都打印了"start..."，然后一起等待 1000ms，再一起输出"end..."。

好了，现在我们给 test() 方法添加 synchronized 关键字，避免多线程并行访问的问题，如代码 6-20 所示。

代码 6-20

```java
public synchronized static void test() {
    System.out.println("start...");
    try {
        Thread.currentThread().sleep(1000);
    } catch (InterruptedException e) {
        // TODO Auto-generated catch block
        e.printStackTrace();
    }
    System.out.println("end...");
}
```

再次运行程序，结果如图 6-19 所示。

```
start...
end...
start...
end...
start...
end...
start...
end...
start...
end...
```

图 6-19

此时的结果是"start..."、"end..."成对输出，因为我们给 test() 方法加了一把锁，线程 1 先到，拿到了这把锁，然后执行 test 的业务逻辑。在整个执行过程中，其他线程到了也只能处于阻塞状态等待，因为它们拿不到锁。只有当线程 1 执行完毕释放了锁，其他线程才可以拿到锁进而执行方法。所以当前实际的执行情况是线程 1 输出 "start..."，等待 1000ms 之后输出 "end..."，紧接着线程 2 输出 "start..."，等待 1000ms 之后输出 "end..."，以此类推。上述代码中 synchronized 修饰的是静态方法，现在用 synchronized 修饰实例方法，如代码 6-21 所示。

代码 6-21

```java
public class SynchronizedTest {
    public static void main(String[] args) {
        for(int i = 0; i < 5;i++) {
            Thread thread = new Thread(new Runnable() {
                @Override
                public void run() {
                    // TODO Auto-generated method stub
                    SynchronizedTest synchronizedTest = new SynchronizedTest();
                    synchronizedTest.test();
                }
            });
            thread.start();
```

```
            }
        }
        public synchronized void test() {
            System.out.println("start...");
            try {
                Thread.currentThread().sleep(1000);
            } catch (InterruptedException e) {
                // TODO Auto-generated catch block
                e.printStackTrace();
            }
            System.out.println("end...");
        }
    }
```

运行结果如图 6-20 所示。

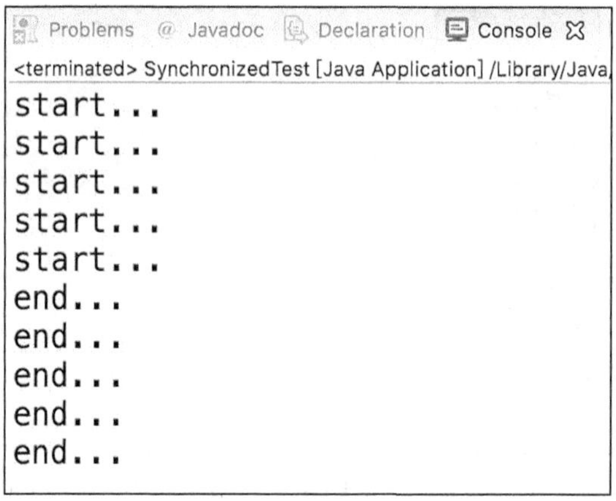

```
start...
start...
start...
start...
start...
end...
end...
end...
end...
end...
```

图 6-20

通过结果可以看到，此时的内置锁并没有为一个线程锁定资源，加锁的效果并没有实现，这是为什么呢？因为我们当前锁定的是一个实例方法，每一个线程都有这样的一个实例方法，相互之间是独立的。即 test 方法并不是被所有线程所共享的，实际的运行情况是每一个线程都获取自己的锁，然后并行访问，相互之间并没有"你运行、我等待"的关系，所以给实例方法添加 synchronized 关键字并不能实现线程同步。

看到这里，大家可能会有疑问了，本章的第一个案例"统计访问量"中，synchronized 修饰的也是实例方法，为什么就可以实现同步呢？因为实例方法中操作的变量 num 是静态的，所以还是多线程在共享资源，线程同步的本质是锁定多个线程所共享的资源。synchronized 还可以修饰代码块，会为代码块加上内置锁，从而实现同步。在静态方法中添加 synchronized 可以同步代码块，具体实现如代码 6-22 所示。

代码 6-22

```
public class SynchronizedTest {
    public static void main(String[] args) {
        for(int i = 0; i < 5;i++) {
            Thread thread = new Thread(new Runnable() {
                @Override
                public void run() {
```

```
            // TODO Auto-generated method stub
            SynchronizedTest.test();
          }
      });
      thread.start();
    }
  }
  public static void test() {
    synchronized (SynchronizedTest.class) {
      System.out.println("start...");
      try {
        Thread.currentThread().sleep(1000);
      } catch (InterruptedException e) {
        // TODO Auto-generated catch block
        e.printStackTrace();
      }
      System.out.println("end...");
    }
  }
}
```

synchronized()内设置需要加锁的资源，静态方法是属于类的方法，不属于任何一个实例对象。所以静态方法中的 synchronized() 只能锁定类，不能锁定实例，this 可以表示当前的一个实例，如图 6-21 所示。

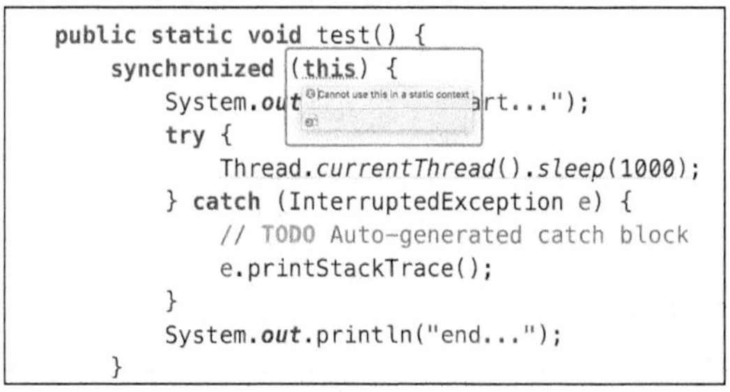

```
public static void test() {
    synchronized (this) {
        System.out ⓘ Cannot use this in a static context rt...");
        try {
            Thread.currentThread().sleep(1000);
        } catch (InterruptedException e) {
            // TODO Auto-generated catch block
            e.printStackTrace();
        }
        System.out.println("end...");
    }
```

图 6-21

程序会自动报错，静态方法中不能使用 this 关键字。同理静态方法中也不能锁定实例变量，只能锁定静态变量，代码 6-22 运行程序结果如图 6-22 所示。

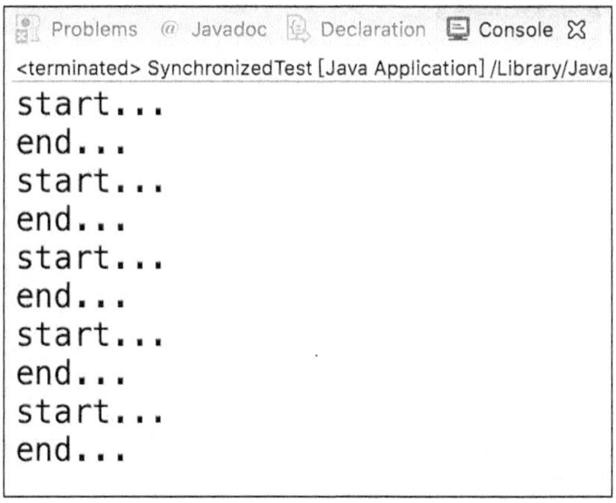

```
Problems  @ Javadoc  Declaration  Console
<terminated> SynchronizedTest [Java Application] /Library/Java
start...
end...
start...
end...
start...
end...
start...
end...
start...
end...
```

图 6-22

在实例方法中也可以添加 synchronized 同步代码块，具体实现如代码 6-23 所示。

代码 6-23

```java
public class SynchronizedTest {
   public static void main(String[] args) {
      for(int i = 0; i < 5;i++) {
         Thread thread = new Thread(new Runnable() {
            @Override
            public void run() {
               // TODO Auto-generated method stub
               SynchronizedTest synchronizedTest = new SynchronizedTest();
               synchronizedTest.test();
            }
         });
         thread.start();
      }
   }
   public void test() {
      synchronized (this) {
         System.out.println("start...");
         try {
            Thread.currentThread().sleep(1000);
         } catch (InterruptedException e) {
            // TODO Auto-generated catch block
            e.printStackTrace();
         }
         System.out.println("end...");
      }
   }
}
```

运行结果如图 6-23 所示。

```
Problems  @ Javadoc  Declaration  Console
<terminated> SynchronizedTest [Java Application] /Library/Java
start...
start...
start...
start...
start...
end...
end...
end...
end...
end...
```

图 6-23

通过结果可以得知没有实现代码同步，原因是 synchronized() 锁定的是 this，即当前的实例对象。每个线程都有一个实例对象，相互独立，并不是共享资源，所以没有实现线程同步，该如何修改呢？只需要锁住共享资源即可，实例对象是每个线程独有的，类则是共享的，所以锁定类即可，如代码 6-24 所示。

代码 6-24

```java
public void test() {
    synchronized (SynchronizedTest.class) {
        System.out.println("start...");
        try {
            Thread.currentThread().sleep(1000);
        } catch (InterruptedException e) {
            // TODO Auto-generated catch block
            e.printStackTrace();
        }
        System.out.println("end...");
    }
}
```

运行结果如图 6-24 所示。

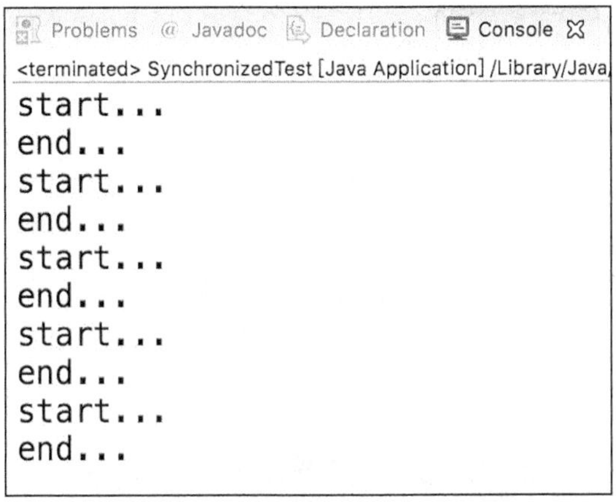

```
start...
end...
start...
end...
start...
end...
start...
end...
start...
end...
```

图 6-24

6.4.2 线程安全的单例模式

单例模式是一种常见的软件设计模式，其核心思想是一个类只有一个实例对象，由多个线程来共享该实例对象资源。在某些系统中，只有一个实例对象资源很重要。例如售票系统，共 1000 张票，分 10 个窗口出售，则这 10 个窗口就必须共享这 1000 张票的实例对象。如果每个窗口都有自己的实例对象，就变成了每个窗口都有 1000 张票可以出售，有悖于真实逻辑。单例模式如何实现呢？核心是共享实例对象，那么就把实例对象定义为静态，如代码 6-25 所示。

代码 6-25

```java
public class SingletonDemo {
    private static SingletonDemo instance;
    private SingletonDemo() {
        System.out.println("SingletonDemo...");
    }
    public static SingletonDemo getInstance() {
        if(instance == null) {
            instance = new SingletonDemo();
```

```
        }
        return instance;
    }
}

public class Test {
    public static void main(String[] args) {
        SingletonDemo instance = SingletonDemo.getInstance();
        SingletonDemo instance2 = SingletonDemo.getInstance();
    }
}
```

运行结果如图 6-25 所示。

图 6-25

你可以看到这里只创建了一个实例对象,也实现了共享,但是这并不是真正的单例模式。我们在写代码的时候一定要考虑到多线程并行访问的情况,现在修改代码实现多线程并行访问,如代码 6-26 所示。

代码 6-26

```
public class Test {
    public static void main(String[] args) {
        new Thread(new Runnable() {
            @Override
            public void run() {
                // TODO Auto-generated method stub
                SingletonDemo instance = SingletonDemo.getInstance();
            }
        }).start();
        //重复上述代码
    }
}
```

运行结果如图 6-26 所示。

图 6-26

这里创建了两个实例对象，原因是线程 1 和线程 2 是并行访问的。线程 1 先来判断 instance==null 是成立的，然后线程 1 来实例化对象。正在此时，实例化对象的操作还没有完成，线程 2 来了。先判断 instance==null 是成立的，于是线程 2 也执行了实例化对象的操作，所以导致实例化了两个对象，如图 6-27 所示。

图 6-27

那么如何进行优化呢？通过加锁实现线程同步即可，我们给 getInstance()方法加一把锁，用 synchronized 来修饰方法，如代码 6-27 所示。

代码 6-27

```
public class SingletonDemo {
   private static SingletonDemo instance;
   private SingletonDemo() {
      System.out.println("SingletonDemo...");
   }
   public synchronized static SingletonDemo getInstance() {
      if(instance == null) {
         instance = new SingletonDemo();
      }
      return instance;
   }
}
```

再次运行程序，结果如图 6-28 所示。

图 6-28

我们说过 synchronized 可以修饰方法，也可以修饰代码块。接下来，我们通过同步代码块的方式来实现单例模式，如代码 6-28 所示。

代码 6-28

```
public class SingletonDemo {
   private volatile static SingletonDemo instance;
   private SingletonDemo() {
      System.out.println("SingletonDemo...");
   }
   public static SingletonDemo getInstance() {
      if(instance == null) {
```

```
            synchronized (SingletonDemo.class) {
                if(instance == null) {
                    instance = new SingletonDemo();
                }
            }
        }
        return instance;
    }
}
```

运行程序，结果如图 6-29 所示。

图 6-29

这里使用了 volatile 关键字修饰 instance，volatile 的作用是可以使内存中的数据对线程可见，这句话是什么意思呢？首先要说明一下 Java 的内存模型，一个线程在访问内存数据时，其实不是拿到该数据本身，而是将该数据复制保存到工作内存中。相当于取出一个副本，对工作内存中的数据进行修改，再保存到主内存中，即主内存对线程是不可见的。所以当线程 1 拿到锁，并锁定整个类之后，就实例化了 instance 对象，即"instance = new SingletonDemo();"。但是此时的 instance 是工作内存中的数据，还需要将工作内存中的数据保存到主内存中。然而锁定的只是实例化的步骤，保存到主内存的步骤没有加锁。所以工作内存中的 instance 完成实例化之后，还未更新到主内存之前就释放了锁。线程 2 立即获取锁，又从主内存复制了一份数据，此时的数据还是 null。线程 2 又在工作内存中完成了一次实例化，然后线程 1 和线程 2 再将它们各自实例化之后的数据保存到主内存中，如图 6-30 所示。

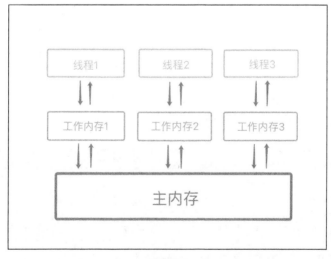

图 6-30

6.4.3 死锁

使用 synchronized 可以实现线程同步，这可以解决多线程并行访问数据带来的安全问题。但是任何事物都没有绝对的好与坏，synchronized 在解决线程安全问题的同时也会带来一个隐患，并且这个隐患是比较严重的，那就是死锁。

先来解释一下死锁的概念，举个生活中的例子，10 个人围一桌吃饭，但是每个人只有一根筷子，要求必须凑齐一双筷子才可以吃菜。也就是说每个人必须要拿到其他人的筷子，但是每个人又都不愿意把自己的筷子让给别人，都在等待其他人主动把筷子贡献出来。这样就形成了一个死局，如果一直保存这种状态，这个饭局就一直僵在这里，没有人可以吃到菜，这就是死锁。如果把每个人看成一个线程，筷子就是线程要获取的资源，现在每个线程都占用一个资源并且不愿意释放，而且任意一个线程想继续执行就必须获取其他线程的资源，那么所有的线程都处于阻塞状态，程序无法向下执行也无法结束。

如何破解死锁呢？唯有某个线程愿意作出让步，贡献出自己的资源给其他线程使用。获取到资源的线程就可以执行自己的业务方法，执行完毕后会释放它所占用的两个资源，其他线程就可以依次获取资源来执行业务方法，问题就迎刃而解了。相当于在那个僵持的饭局上，有一个人愿意作出牺牲，把自己的筷子让给身边的人，这样他身边的人就可以吃饭了，待他吃饱之后，会把他的筷子贡献出来，这样所有人都可以依次吃到饭了。我们通过一段代码来演示死锁的情况，如代码 6-29 所示。

代码 6-29

```java
public class Chopsticks {
}

public class DeadLockRunnable implements Runnable{
   public int num;
   private static Chopsticks chopsticks1 = new Chopsticks();
   private static Chopsticks chopsticks2 = new Chopsticks();
   @Override
   public void run() {
      // TODO Auto-generated method stub
      if(num == 1){
         System.out.println(Thread.currentThread().getName()+"获取到 chopsticks1,等待获取 chopsticks2");
         synchronized (chopsticks1) {
            try {
               Thread.sleep(100);
            } catch (InterruptedException e) {
               // TODO Auto-generated catch block
               e.printStackTrace();
            }
            synchronized (chopsticks2) {
               System.out.println(Thread.currentThread().getName()+"用餐完毕");
            }
         }
      }
      if(num == 2){
         System.out.println(Thread.currentThread().getName()+"获取到 chopsticks2,等待获取 chopsticks1");
```

```java
            synchronized (chopsticks2) {
                try {
                    Thread.sleep(100);
                } catch (InterruptedException e) {
                    // TODO Auto-generated catch block
                    e.printStackTrace();
                }
                synchronized (chopsticks1) {
                    System.out.println(Thread.currentThread().getName()+"用餐完毕");
                }
            }
        }
    }
}

public class DeadLockTest {
    public static void main(String[] args) {
        DeadLockRunnable deadLockRunnable1 = new DeadLockRunnable();
        deadLockRunnable1.num = 1;
        DeadLockRunnable deadLockRunnable2 = new DeadLockRunnable();
        deadLockRunnable2.num = 2;
        new Thread(deadLockRunnable1,"张三").start();
        new Thread(deadLockRunnable2,"李四").start();
    }
}
```

运行程序，结果如图 6-31 所示。

图 6-31

张三获取了资源 chopsticks1，必须同时获取 chopsticks2 才能完成用餐，但是 chopsticks2 被李四占用，李四只有完成用餐才能释放 chopsticks2，但是李四完成用餐的必要条件是获取 chopsticks1，所以形成了一种互斥的关系，这就是死锁。可以对代码进行修改，不要让两个线程并行访问，先启动代表张三的线程，休眠 2000ms，待张三完成用餐释放资源之后再启动代表李四的线程，如代码 6-30 所示。

代码 6-30

```java
public class DeadLockTest {
    public static void main(String[] args) {
        DeadLockRunnable deadLockRunnable1 = new DeadLockRunnable();
        deadLockRunnable1.num = 1;
        DeadLockRunnable deadLockRunnable2 = new DeadLockRunnable();
        deadLockRunnable2.num = 2;
        new Thread(deadLockRunnable1,"张三").start();
        try {
```

```
            Thread.currentThread().sleep(2000);
        } catch (InterruptedException e) {
        }
        new Thread(deadLockRunnable2,"李四").start();
    }
}
```

运行程序，结果如图 6-32 所示。

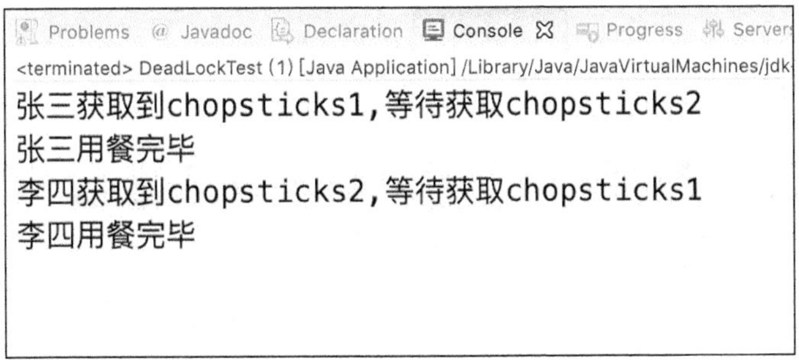
Problems
Javadoc
Declaration
Console
Progress
Server
<terminated> DeadLockTest (1) [Java Application] /Library/Java/JavaVirtualMachines/jdk
张三获取到chopsticks1,等待获取chopsticks2
张三用餐完毕
李四获取到chopsticks2,等待获取chopsticks1
李四用餐完毕

图 6-32

这样就可以解决死锁导致的问题，实际上死锁是一种错误，在实际开发中需要注意避免这种错误的出现。

6.4.4 重入锁

重入锁（ReentrantLock）是对 synchronized 的升级，synchronized 是通过 JVM 实现的，ReentrantLock 是通过 JDK 实现的，重入锁有什么特点呢？顾名思义，重入锁指可以给同一个资源添加多个锁，并且解锁的方式与 synchronized 也有不同。synchronized 的锁是线程执行完毕之后会自动释放的，ReentrantLock 的锁必须手动释放，可以通过 ReentrantLock 实现访问量统计，如代码 6-31 所示。

代码 6-31

```
public class Account implements Runnable{
    private static int num;
    private ReentrantLock reentrantLock = new ReentrantLock();
    @Override
    public void run() {
        // TODO Auto-generated method stub
        reentrantLock.lock();
        num++;
        System.out.println(Thread.currentThread().getName()+"是当前的第"+num+"位访客");
        reentrantLock.unlock();
    }
}
```

首先需要实例化 ReentrantLock 的成员变量，在业务方法中需要加锁的地方直接调用对象的 lock()方法即可，同理需要解锁的地方直接调用对象的 unlock()方法即可，运行程序，结果如图 6-33 所示。

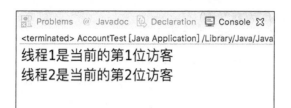

图 6-33

效果与 synchronized 一样,在此基础之上可以添加多把锁,只需要多次调用 lock() 方法即可,如代码 6-32 所示。

代码 6-32

```java
public class Account implements Runnable{
    private static int num;
    private ReentrantLock reentrantLock = new ReentrantLock();
    @Override
    public void run() {
        // TODO Auto-generated method stub
        reentrantLock.lock();
        reentrantLock.lock();
        num++;
        System.out.println(Thread.currentThread().getName()+"是当前的第"+num+"位访客");
        reentrantLock.unlock();
        reentrantLock.unlock();
    }
}
```

我们说过 ReentrantLock 需要手动解锁,如果我们只加锁而不手动解锁,如代码 6-33 所示。

代码 6-33

```java
public class Account implements Runnable{
    private static int num;
    private ReentrantLock reentrantLock = new ReentrantLock();
    @Override
    public void run() {
        // TODO Auto-generated method stub
        reentrantLock.lock();
        reentrantLock.lock();
        num++;
        System.out.println(Thread.currentThread().getName()+"是当前的第"+num+"位访客");
        reentrantLock.unlock();
    }
}
```

运行程序,结果如图 6-34 所示。

可以看到这里只输出了第 1 位访客信息,然后程序就静止了,一直处于运行状态但是不会自动结束。原因就是线程 1 进来,加了两把锁。执行完线程 1 的输出语句之后,只释放了一把锁。线程 2 永远无法获得第 2 把锁,就一直处于阻塞状态,程序也就一直卡在这里了,所以使用重入锁的时候需要注意加了几把锁就必须释放几把锁。

图 6-34

ReentrantLock 除了可重入之外，还有一个可中断的特点，可中断是指某个线程在等待获取锁的过程中可主动终止线程，通过调用对象的 lockInterruptibly() 来实现。如这样一个场景，线程 1 和线程 2 并行访问某个方法，该方法执行完成需要 5000ms。肯定会有一个线程先拿到锁，然后另外一个线程就进入阻塞状态。现在设置当线程启动 1000ms 之后，没有拿到锁的线程自动中断，具体实现如代码 6-34 所示。

代码 6-34

```java
public class StopLock implements Runnable {
public ReentrantLock reentrantLock = new ReentrantLock();
    @Override
    public void run() {
      // TODO Auto-generated method stub
      try {
        reentrantLock.lockInterruptibly();
        System.out.println(Thread.currentThread().getName()+" get lock");
        Thread.currentThread().sleep(5000);
      } catch (Exception e) {
        // TODO: handle exception
        e.printStackTrace();
      } finally {
        reentrantLock.unlock();
      }
    }
}


public class Test {
    public static void main(String[] args) {
       StopLock lock = new StopLock();
       Thread t1 = new Thread(lock, "线程1");
       t1.start();
       Thread t2 = new Thread(lock, "线程2");
       t2.start();
       try {
          Thread.currentThread().sleep(1000);
          t2.interrupt();
       } catch (InterruptedException e) {
          // TODO Auto-generated catch block
          e.printStackTrace();
       }
    }
}
```

运行程序，结果如图 6-35 所示。

```
线程1 get lock
java.lang.InterruptedException
        at java.base/java.util
        at java.base/java.util
        at java.base/java.util
        at thread.reentrantloc
        at java.base/java.lang
```

图 6-35

ReentrantLock 还具备限时性的特点，指可以判断某个线程在一定的时间内能否获取锁，通过调用 tryLock(long timeout, TimeUnit unit)方法来实现。其中 timeout 指时间数值，unit 指时间单位，返回值为 boolean 类型，true 表示在该时间段内获取了锁，false 表示在该时间段内没有获取锁，具体实现如代码 6-35 所示。

代码 6-35

```java
public class TimeLock implements Runnable{
    public ReentrantLock reentrantLock = new ReentrantLock();
    @Override
    public void run() {
        // TODO Auto-generated method stub
        try {
            if(reentrantLock.tryLock(3, TimeUnit.SECONDS)) {
                System.out.println(Thread.currentThread().getName()+"get lock");
                Thread.currentThread().sleep(5000);
            }else {
                System.out.println(Thread.currentThread().getName()+"not lock");
            }
        } catch (InterruptedException e) {
            // TODO Auto-generated catch block
            e.printStackTrace();
        } finally {
            if(reentrantLock.isHeldByCurrentThread()) {
                reentrantLock.unlock();
            }
        }
    }
}


public class Test {
    public static void main(String[] args) {
        TimeLock lock = new TimeLock();
        new Thread(lock,"线程1").start();
        new Thread(lock,"线程2").start();
    }
}
```

线程 1 和线程 2 并行访问，业务方法的执行需要 5000ms。reentrantLock.tryLock(3, TimeUnit.SECONDS)表示如果线程启动之后的 3s 内该线程没有拿到锁，则返回 false，反之返回 true。很显然 5000ms 大于 3s，会有一个线程是肯定拿不到锁的，运行程序，结果如图 6-36 所示。

图 6-36

线程 1 拿到了锁，线程 2 没有拿到锁，因为它的等待时间为 3s，而线程 1 从拿到锁到释放锁需要 5s，现在对程序进行修改，如代码 6-36 所示。

代码 6-36

```java
public class TimeLock implements Runnable{
    public ReentrantLock reentrantLock = new ReentrantLock();
    @Override
    public void run() {
        // TODO Auto-generated method stub
        try {
            if(reentrantLock.tryLock(6, TimeUnit.SECONDS)) {
                System.out.println(Thread.currentThread().getName()+"get lock");
                Thread.currentThread().sleep(5000);
            }else {
                System.out.println(Thread.currentThread().getName()+"not lock");
            }
        } catch (InterruptedException e) {
            // TODO Auto-generated catch block
            e.printStackTrace();
        } finally {
            if(reentrantLock.isHeldByCurrentThread()) {
                reentrantLock.unlock();
            }
        }
    }
}
```

reentrantLock.tryLock(6, TimeUnit.SECONDS)表示如果线程启动之后的 6s 内该线程没有拿到锁，则返回 false，反之返回 true，运行程序，结果如图 6-37 所示。

图 6-37

线程 1 拿到了锁，线程 2 也拿到了锁，因为等待时间为 6s，线程 1 从获取锁到释放锁的时间是 5s。

6.4.5 生产者消费者模式

生产者消费者意为在一个生产环境中,生产者和消费者在同一个时间段内共享同一块缓冲区。生产者负责向缓冲区中添加数据,消费者负责从缓冲区中取出数据。以生产汉堡和消费汉堡为例来实现生产者消费者模式,具体实现如代码 6-37 所示。

代码 6-37

```java
//汉堡类：
public class Hamburger {
    private int id;
    //getter、setter 方法

    public Hamburger(int id){
        this.id = id;
    }
    @Override
    public String toString() {
        return "Hamburger [id=" + id + "]";
    }
}

//装汉堡的容器类：
public class Container {
    public Hamburger[] array = new Hamburger[6];
    public int index = 0;
    //向容器中添加汉堡
    public synchronized void push(Hamburger hamburger){
        while(index == array.length){
            try {
                this.wait();
            } catch (InterruptedException e) {
                // TODO Auto-generated catch block
                e.printStackTrace();
            }
        }
        this.notify();
        array[index] = hamburger;
        index++;
        System.out.println("生产了一个汉堡："+hamburger);
    }
    //从容器中取出汉堡
    public synchronized Hamburger pop(){
        while(index == 0){
            try {
                this.wait();
            } catch (InterruptedException e) {
                // TODO Auto-generated catch block
                e.printStackTrace();
            }
        }
```

```java
            this.notify();
            index--;
            System.out.println("消费了一个汉堡："+array[index]);
            return array[index];
        }
    }

//生产者类：
public class Producer implements Runnable{
    private Container container = null;
    public Producer(Container container){
        this.container = container;
    }
    @Override
    public void run() {
        // TODO Auto-generated method stub
        for(int i = 0; i < 30; i++){
            Hamburger hamburger = new Hamburger(i);
            this.container.push(hamburger);
            try {
                Thread.currentThread().sleep(1000);
            } catch (InterruptedException e) {
                // TODO Auto-generated catch block
                e.printStackTrace();
            }
        }
    }
}


//消费者类：
public class Consumer implements Runnable{
    private Container container = null;
    public Consumer(Container container){
        this.container = container;
    }
    @Override
    public void run() {
        // TODO Auto-generated method stub
        for(int i = 0; i < 30;i++){
            this.container.pop();
            try {
                Thread.currentThread().sleep(1000);
            } catch (InterruptedException e) {
                // TODO Auto-generated catch block
                e.printStackTrace();
            }
        }
    }
}

//测试类：
public class Test {
    public static void main(String[] args) {
        Container container = new Container();
```

```
        Producer producer = new Producer(container);
        Consumer consumer = new Consumer(container);
        new Thread(producer).start();
        new Thread(producer).start();
        new Thread(consumer).start();
        new Thread(consumer).start();
        new Thread(consumer).start();
    }
}
```

运行程序，结果如图 6-38 所示。

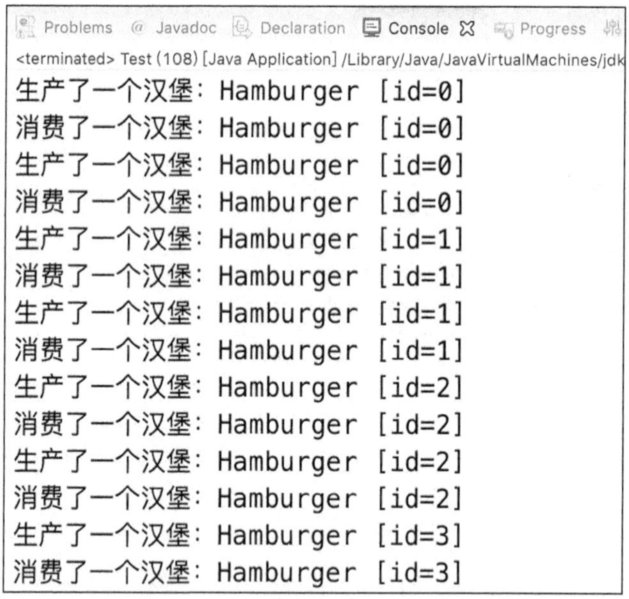

```
生产了一个汉堡: Hamburger [id=0]
消费了一个汉堡: Hamburger [id=0]
生产了一个汉堡: Hamburger [id=0]
消费了一个汉堡: Hamburger [id=0]
生产了一个汉堡: Hamburger [id=1]
消费了一个汉堡: Hamburger [id=1]
生产了一个汉堡: Hamburger [id=1]
消费了一个汉堡: Hamburger [id=1]
生产了一个汉堡: Hamburger [id=2]
消费了一个汉堡: Hamburger [id=2]
生产了一个汉堡: Hamburger [id=2]
消费了一个汉堡: Hamburger [id=2]
生产了一个汉堡: Hamburger [id=3]
消费了一个汉堡: Hamburger [id=3]
```

图 6-38

6.5 综合练习

一场球赛的球票分 3 个窗口出售，共 15 张票，请用多线程模拟 3 个窗口的售票情况，具体实现如代码 6-38 所示。

代码 6-38

```java
public class TicketRunnable implements Runnable{
    //剩余球票数
    public int surpluCount = 15;
    //已售出球票数
    public int outCount = 0;
    @Override
    public void run() {
        // TODO Auto-generated method stub
        while(surpluCount > 0){
            try {
```

```java
                Thread.sleep(500);
            } catch (InterruptedException e) {
            }
            if(surpluCount == 0){
                return;
            }
            synchronized (TicketRunnable.class) {
                surpluCount--;
                outCount++;
                if(surpluCount == 0) {
                    System.out.println(Thread.currentThread().getName()+"售出第"+outCount+"张球票，球票已售罄");
                }else {
                    System.out.println(Thread.currentThread().getName()+"售出第"+outCount+"张球票，剩余"+surpluCount+"张球票");
                }
            }
        }
    }
}

public class TicketTest {
    public static void main(String[] args) {
        TicketRunnable ticketRunnable = new TicketRunnable();
        new Thread(ticketRunnable,"窗口A").start();
        new Thread(ticketRunnable,"窗口B").start();
        new Thread(ticketRunnable,"窗口C").start();
    }
}
```

程序的运行结果如图 6-39 所示。

```
Problems  @ Javadoc  Declaration  Console  Progress  Se
<terminated> TicketTest [Java Application] /Library/Java/JavaVirtualMachines/jdk-10
窗口A售出第1张球票，剩余14张球票
窗口C售出第2张球票，剩余13张球票
窗口B售出第3张球票，剩余12张球票
窗口C售出第4张球票，剩余11张球票
窗口A售出第5张球票，剩余10张球票
窗口B售出第6张球票，剩余9张球票
窗口C售出第7张球票，剩余8张球票
窗口B售出第8张球票，剩余7张球票
窗口A售出第9张球票，剩余6张球票
窗口B售出第10张球票，剩余5张球票
窗口C售出第11张球票，剩余4张球票
窗口A售出第12张球票，剩余3张球票
窗口B售出第13张球票，剩余2张球票
窗口A售出第14张球票，剩余1张球票
窗口C售出第15张球票，球票已售罄
```

图 6-39

6.6 小结

本章为大家讲解了多线程的相关知识，如果说我们之前写的代码只能算是测试案例的话，加入多线程机制后的程序才有点正式项目的意思。因为任何一个程序都不可能是一个用户来访问，一定是多个客户端同时来访问一台服务器，是多对一的关系。多个客户端同时访问一个服务器会不会出问题呢？比如一个客服人员同时接听10个用户电话，很有可能出现数据错误，例如把应该给张三的信息传给了李四。我们在程序中要避免这类问题的出现就可以使用线程同步，将一个用户的请求封装成一个线程任务，由系统来调度协调资源，使多个任务可以并行访问并且保证数据不会出错。多线程的知识很重要，需要掌握的知识点较多，如线程的概念、Java中如何创建线程、线程调度、线程同步等。

第 7 章 集合框架

> 集合框架是 Java 很重要的一个知识点，在实际开发中使用的频率较高，几乎可以说是 Java 程序中必备的模块，掌握集合框架的体系结构，了解每个组成部分的特点及作用，可以帮助我们编写高质量的程序。

7.1 集合的概念

首先思考一个问题，我们为什么要使用集合？使用集合会给我们带来哪些好处？带着这样的问题我们来看一个需求，程序中通常需要创建多个对象，同时这些对象的个数以及数据类型在程序编写阶段是无法确定的，需要根据程序的具体运行情况来决定对象的个数及数据类型。那么问题来了，我们怎么保存这些对象呢？第一个我们会想到数组，但很显然数组是无法满足这个需求的，首先数组的长度是固定的，后期无法进行扩容。再有数组的数据类型也是固定的，无法同时存放多个不同数据类型的数据，所以数组是无法解决上述问题的。这时集合就应运而生了，集合可以简单理解为一个长度可以改变，可以保存任意数据类型的动态数组。

集合本身也是数据结构的基本概念之一，我们这里说的集合是 Java 语言对这种数据结构的具体实现。在 Java 中，集合不是由一个类来完成的，而是由一组接口和类构成了一个框架体系。大致可分为 3 层，最上层是一组接口，继而是接口的实现类，接下来是对集合进行各种操作的工具类，集合框架中常用接口的具体描述如表 7-1 所示。

表 7-1

接口	描述
Collection	集合框架最基本的接口，一个 Collection 存储一组无序、不唯一的对象，一般不直接使用该接口
List	Collection 的子接口，存储一组有序、不唯一的对象，常用的接口之一
Set	Collection 的子接口，存储一组无序、唯一的对象
Map	独立于 Collection 的另外一个接口，存储一组键值对象，提供键到值的映射
Iterator	输出集合元素的接口，一般适用于无序集合，从前到后单向输出
ListIterator	Iterator 的子接口，可以双向输出集合中的元素
Enumeration	传统的输出接口，已被 Iterator 所取代

续表

接口	描述
SortedSet	Set 的子接口，可对集合中的元素进行排序
SortedMap	Map 的子接口，可对集合中的键值元素进行排序
Queue	队列接口，此接口的子类可实现队列操作
Map.Entry	Map 的内部接口，描述 Map 中的一个键值对元素

接下来介绍常用接口及其实现类。

7.2 Collection 接口

7.2.1 Collection 接口的定义

Collection 是集合框架中最基础的父接口，可以存储一组无序，不唯一的对象。一般不直接使用该接口，也不能被实例化，只是用来提供规范定义，Collection 接口的定义如图 7-1 所示。

```
public interface Collection<E> extends Iterable<E> {
    // Query Operations
```

图 7-1

可以看到 Collection 是 Iterable 的子接口，Collection 和 Iterable 后面的<E>表示它们都使用了泛型的定义，泛型是指在操作集合时需要指定具体的数据类型，这样可以保证数据的安全性。在后面的章节我们会详细讲解泛型，Collection 接口常用方法的描述如表 7-2 所示。

表 7-2

方法	描述
int size()	获取集合长度
boolean isEmpty()	判断集合是否为空
boolean contains(Object o)	判断集合中是否存在某个对象
Iterator<E> iterator()	实例化 Iterator 接口，遍历集合
Object[] toArray()	将集合转换为一个 Object 类型的对象数组
<T> T[] toArray(T[] a)	将集合转换为一个指定数据类型的对象数组
boolean add(E e)	向集合中添加元素
boolean remove(Object o)	从集合中移除元素
boolean containsAll(Collection<?> c)	判断集合中是否存在某个集合的所有元素
boolean addAll(Collection<? extends E> c)	向集合中添加某个集合的所有元素
boolean removeAll(Collection<?> c)	从集合中移除某个集合的所有元素
default boolean removeIf(Predicate<? super E> filter)	从集合中移除满足给定条件的集合的所有元素
boolean retainAll(Collection<?> c)	对集合进行操作，只保留包含在目标集合中的元素

续表

方法	描述
void clear()	清除集合中的所有元素
boolean equals(Object o)	比较两个集合是否相等
int hashCode()	获取集合的散列值
default Spliterator<E> spliterator()	将集合转换为一个指定数据类型的并行迭代器
default Stream<E> stream()	将集合转换为一个流
default Stream<E> parallelStream()	将集合转换为一个可并行的流

7.2.2 Collection 的子接口

Collection 作为集合的基本接口，在实际开发中一般不直接使用，而是使用其子接口进行开发，Collection 主要的子接口如下。

- List：存放有序，不唯一的元素。
- Set：存放无序，唯一的元素。
- Queue：队列接口。

7.3 List 接口

7.3.1 List 接口的定义

List 是 Collection 的常用子接口，可以存储一组有序、不唯一的对象，List 接口的定义如图 7-2 所示。

```
public interface List<E> extends Collection<E> {
    // Query Operations
```

图 7-2

List 接口在继承 Collection 接口的基础上进行了扩展，常用的扩展方法如表 7-3 所示。

表 7-3

方法	描述
E get(int index)	通过下标获取集合中指定位置的元素
E set(int index, E element)	替换集合中指定位置的元素
void add(int index, E element)	向集合中的指定位置添加元素
E remove(int index)	通过下标删除集合中指定位置的元素
int indexOf(Object o)	查找某个对象在集合中的位置
int lastIndexOf(Object o)	从后向前查找某个对象在集合中的位置
ListIterator<E> listIterator()	实例化 ListIterator 接口
List<E> subList(int fromIndex, int toIndex)	获取集合中的子集合

7.3.2　List 接口的实现类

　　ArrayList 是开发中经常使用到的实现类，实现了长度可变的数组。可以在内存中分配连续的空间，底层是基于索引的数据结构，所以访问元素效率较高。使用索引查询元素可快速访问到对应的元素，但缺点是若添加或者删除元素，需要移动兄弟元素的位置，效率较低，相关的定义如图 7-3 所示。

```
public class ArrayList<E> extends AbstractList<E>
        implements List<E>, RandomAccess, Cloneable, java.io.Serializable
```

图 7-3

　　ArrayList 的使用如代码 7-1 所示。

代码 7-1

```java
public class Test {
    public static void main(String[] args) {
        ArrayList list = new ArrayList();
        list.add("Hello");
        list.add("World");
        list.add("JavaSE");
        list.add("JavaME");
        list.add("JavaEE");
        System.out.println("list:"+list);
        System.out.println("list 的长度："+list.size());
        System.out.println("list 是否包含 Java："+list.contains("Java"));
        Iterator iter = list.iterator();
        while(iter.hasNext()) {
            System.out.print(iter.next()+",");
        }
        for(int i = 0; i < list.size(); i++) {
            System.out.print(list.get(i)+",");
        }
        list.remove("Hello");
        list.remove(0);
        for(int i = 0; i < list.size(); i++) {
            System.out.print(list.get(i)+",");
        }
        list.add(1, "我爱学 Java");
        for(int i = 0; i < list.size(); i++) {
            System.out.print(list.get(i)+",");
        }
        list.set(2, "ArrayList 详解");
        for(int i = 0; i < list.size(); i++) {
            System.out.print(list.get(i)+",");
        }
        System.out.println("JavaEE 在集合中的下标："+list.indexOf("JavaEE"));
        List list2 = list.subList(1, 3);
        System.out.println(list2);
    }
}
```

运行结果如图 7-4 所示。

```
list:[Hello, World, JavaSE, JavaME, JavaEE]
list的长度: 5
list是否包含Java: false
Hello,World,JavaSE,JavaME,JavaEE,
Hello,World,JavaSE,JavaME,JavaEE,
----------删除元素----------
JavaSE,JavaME,JavaEE,
----------添加元素----------
JavaSE,我爱学Java,JavaME,JavaEE,
----------替换元素----------
JavaSE,我爱学Java,ArrayList详解,JavaEE,
JavaEE在集合中的下标: 3
----------截取集合----------
[我爱学Java, ArrayList详解]
```

图 7-4

Vector 是一个早期的 List 实现类，用法基本与 ArrayList 一致，定义如图 7-5 所示。

```
public class Vector<E>
    extends AbstractList<E>
    implements List<E>, RandomAccess, Cloneable, java.io.Serializable
{
```

图 7-5

Vector 的使用如代码 7-2 所示。

代码 7-2

```java
public class Test {
    public static void main(String[] args) {
        Vector vector = new Vector();
        vector.addElement("Hello");
        vector.add("Java");
        for(int i = 0; i < vector.size(); i++) {
            System.out.println(vector.get(i));
        }
    }
}
```

运行结果如图 7-6 所示。

图 7-6

Stack 是 Vector 的子类，实现了一个"后进先出"的栈，定义如图 7-7 所示。

```
public
class Stack<E> extends Vector<E> {
```

图 7-7

Stack 的使用如代码 7-3 所示。

代码 7-3

```
public class Test {
   public static void main(String[] args) {
      Stack stack = new Stack();
      stack.push("Hello");
      stack.push("JavaSE");
      stack.push("JavaME");
      stack.push("JavaEE");
      System.out.println(stack);
      for (int i = 0; i < stack.size(); i++) {
         System.out.print(stack.get(i)+",");
      }
      System.out.println("栈顶元素: "+stack.peek());
      for (int i = 0; i < stack.size(); i++) {
         System.out.print(stack.get(i)+",");
      }
      System.out.println("栈顶元素: "+stack.pop());
      for (int i = 0; i < stack.size(); i++) {
         System.out.print(stack.get(i)+",");
      }
   }
}
```

运行结果如图 7-8 所示。

```
Problems  @ Javadoc  Declaration  Console  Progress
<terminated> Test (62) [Java Application] /Library/Java/JavaVirtualMachines/
[Hello, JavaSE, JavaME, JavaEE]
Hello,JavaSE,JavaME,JavaEE,
栈顶元素: JavaEE
Hello,JavaSE,JavaME,JavaEE,
栈顶元素: JavaEE
Hello,JavaSE,JavaME,
```

图 7-8

LinkedList 实现了一个“先进先出”的队列，采用链表的形式存储分散的内存空间。元素和元素之间通过存储彼此的位置信息来形成连接关系，通过位置信息找到前后节点的关系。所以添加或删除元素的效率更高，因为只需要修改前后节点信息即可。LinkedList 不但要保存每个节点的数据，还需要保存前后节点的位置信息，所以它需要更多的内存空间。正因为如此，在索引检索时会很慢，需要从第一个元素开始遍历，查询元素效率低是其缺点，定义如图 7-9 所示。

```
public class LinkedList<E>
    extends AbstractSequentialList<E>
    implements List<E>, Deque<E>, Cloneable, java.io.Serializable
```

图 7-9

LinkedList 的使用如代码 7-4 所示。

代码 7-4

```
public class Test {
   public static void main(String[] args) {
      LinkedList list = new LinkedList();
      list.add("Hello");
      list.add("World");
      list.add("Java");
      System.out.println(list);
      list.offer("JavaSE");
      list.addLast("JavaSE2");
      System.out.println(list);
      list.push("JavaME");
      list.addFirst("JavaME2");
      System.out.println(list);
      System.out.println("第一个元素："+list.peekFirst());
      System.out.println("访问第一个元素之后的集合："+list);
      System.out.println("最后一个元素："+list.peekLast());
      System.out.println("访问最后一个元素之后的集合："+list);
      System.out.println(list.pop());
      System.out.println("取出第一个元素之后的集合："+list);
      System.out.println(list.pollLast());
      System.out.println("取出最后一个元素之后的集合："+list);
   }
}
```

运行结果如图 7-10 所示。

```
Problems  Javadoc  Declaration  Console  Progress  Servers  Debug
<terminated> Test (62) [Java Application] /Library/Java/JavaVirtualMachines/jdk-10.0.1.jdk/Contents/Home/bin/java (2018年9月16日 下午11:15:32)
[Hello, World, Java]
[Hello, World, Java, JavaSE, JavaSE2]
[JavaME2, JavaME, Hello, World, Java, JavaSE, JavaSE2]
第一个元素：JavaME2
访问第一个元素之后的集合：[JavaME2, JavaME, Hello, World, Java, JavaSE, JavaSE2]
最后一个元素：JavaSE2
访问最后一个元素之后的集合：[JavaME2, JavaME, Hello, World, Java, JavaSE, JavaSE2]
JavaME2
取出第一个元素之后的集合：[JavaME, Hello, World, Java, JavaSE, JavaSE2]
JavaSE2
取出最后一个元素之后的集合：[JavaME, Hello, World, Java, JavaSE]
```

图 7-10

LinkedList 和 Stack 都有 pop()方法，都是取出集合的第一个元素，但是可以看到两者的顺序恰好是相反的。Stack 采用的是“后进先出”的方式，是栈的形式；LinkedList 采用的是“先进先出”的方式，是队列的形式。因为 LinkedList 定义时实现了 Deque 接口，而 Deque 是 Queue 的子接口，Queue 又继承自 Collection，它在底层实现了队列的数据结构，定义如图 7-11 所示。

```
public interface Queue<E> extends Collection<E> {
    /**
```

图 7-11

在实际开发中，不能直接实例化 Queue 来完成操作，需要实例化其实现类。Queue 的实现类是 AbstractQueue，同时 AbstractQueue 又是一个抽象类，开发中需要对其子类 PriorityQueue 进行实例化，AbstractQueue 的定义如图 7-12 所示。

```
public abstract class AbstractQueue<E>
    extends AbstractCollection<E>
    implements Queue<E> {
```

图 7-12

PriorityQueue 的定义如图 7-13 所示。

```
public class PriorityQueue<E> extends AbstractQueue<E>
    implements java.io.Serializable {
```

图 7-13

PriorityQueue 在使用时需要注意，添加到该队列中的数据必须是有序的，即对象具备排序的功能，我们来演示一个错误例子，如代码 7-5 所示。

代码 7-5

```
public class Test {
   public static void main(String[] args) {
      PriorityQueue queue = new PriorityQueue();
      queue.add(new A(2));
      queue.add(new A(1));
      System.out.println(queue);
   }
}

class A {
   private int num;
   public A(int num) {
      this.num = num;
   }
}
```

运行结果如图 7-14 所示。

```
Problems  Javadoc  Declaration  Console  Progress  Servers  Debug
<terminated> Test (62) [Java Application] /Library/Java/JavaVirtualMachines/jdk-10.0.1.jdk/Contents/Home/bin/java (2018年9月16日 下午11:44:09)
ng.ClassCastException: com.southwind.collection.A cannot be cast to java.base/java.lang.Comparable
orityQueue.siftUpComparable(PriorityQueue.java:653)
orityQueue.siftUp(PriorityQueue.java:648)
orityQueue.offer(PriorityQueue.java:342)
orityQueue.add(PriorityQueue.java:323)
n.Test.main(Test.java:12)
```

图 7-14

错误原因是 A 的实例化对象无法进行排序，解决方法是让 A 实现 Comparable 接口，并且重写 compareTo 方法，再来完成具体的比较逻辑，如代码 7-6 所示。

代码 7-6

```
class A implements Comparable {
```

```
……
@Override
public int compareTo(Object o) {
    // TODO Auto-generated method stub
    /**
     * A.compareTo(B)
     * 返回值：
     * 1 表示 A 大于 B
     * 0 表示 A 等于 B
     * -1 表示 A 小于 B
     */
    A a = (A) o;
    if(this.num > a.num) {
        return 1;
    }else if(this.num == a.num) {
        return 0;
    }else {
        return -1;
    }
}
@Override
public String toString() {
    return "A [num=" + num + "]";
}
}
```

再次运行程序，结果如图 7-15 所示。

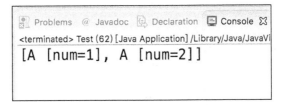

图 7-15

可以看到 PriorityQueue 会对元素进行比较，并按照升序进行排列，即自然排序。同时，PriorityQueue 不允许添加 null 元素。

7.4 Set 接口

7.4.1 Set 接口的定义

Set 是 Collection 的子接口，Set 接口以散列的形式存储数据，所以元素没有顺序。可以存储一组无序且唯一的对象，Set 接口的定义如图 7-16 所示。

```
public interface Set<E> extends Collection<E> {
    // Query Operations
```

图 7-16

实际开发中也不能直接实例化 Set，需要对其实现类进行实例化再完成业务操作。Set 的常用实现类主要有 HashSet、LinkedHashSet、TreeSet。

7.4.2 Set 接口的实现类

HashSet 是开发中经常使用到的实现类，存储一组无序且唯一的对象。这里的无序是指元素的存储顺序和遍历顺序不一致，定义如图 7-17 所示。

```
public class HashSet<E>
    extends AbstractSet<E>
    implements Set<E>, Cloneable, java.io.Serializable
```

图 7-17

HashSet 的使用如代码 7-7 所示。

代码 7-7

```
public class Test {
    public static void main(String[] args) {
        HashSet hashSet = new HashSet();
        hashSet.add("Hello");
        hashSet.add("World");
        hashSet.add("Java");
        hashSet.add("Hello");
        System.out.println("hashSet的长度："+hashSet.size());
        System.out.println("遍历hashSet");
        Iterator iterator = hashSet.iterator();
        while(iterator.hasNext()) {
            System.out.print(iterator.next()+",");
        }
        hashSet.remove("World");
        System.out.println("删除之后遍历hashSet");
        iterator = hashSet.iterator();
        while(iterator.hasNext()) {
            System.out.print(iterator.next()+",");
        }
    }
}
```

运行结果如图 7-18 所示。

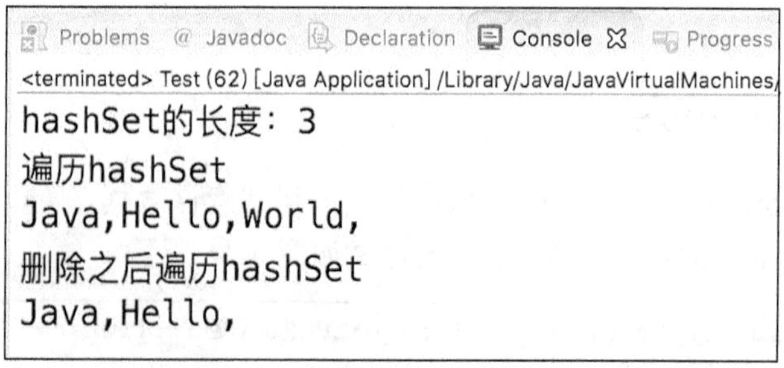

```
hashSet的长度：3
遍历hashSet
Java,Hello,World,
删除之后遍历hashSet
Java,Hello,
```

图 7-18

LinkedHashSet 是 Set 的另外一个子接口，可以存储一组有序且唯一的元素。这里的有序是指元素的存储顺序和遍历顺序是一致的，定义如图 7-19 所示。

```
public class LinkedHashSet<E>
    extends HashSet<E>
    implements Set<E>, Cloneable, java.io.Serializable {
```

图 7-19

LinkedHashSet 的使用如代码 7-8 所示。

代码 7-8

```java
public class Test {
   public static void main(String[] args) {
      LinkedHashSet linkedHashSet = new LinkedHashSet();
      linkedHashSet.add("Hello");
      linkedHashSet.add("World");
      linkedHashSet.add("Java");
      linkedHashSet.add("Hello");
      System.out.println("linkedHashSet的长度："+linkedHashSet.size());
      System.out.println("遍历 linkedHashSet");
      Iterator iterator = linkedHashSet.iterator();
      while(iterator.hasNext()) {
         System.out.print(iterator.next()+",");
      }
      linkedHashSet.remove("World");
      System.out.println("删除之后遍历 linkedHashSet");
      iterator = linkedHashSet.iterator();
      while(iterator.hasNext()) {
         System.out.print(iterator.next()+",");
      }
   }
}
```

运行结果如图 7-20 所示。

```
linkedHashSet的长度：3
遍历linkedHashSet
Hello,World,Java,
删除之后遍历linkedHashSet
Hello,Java,
```

图 7-20

我们对集合执行了添加两个"Hello"的操作，但是只保存了一个。这是因为 LinkedHashSet 集合的元素是唯一的，即不能出现两个相等的元素。字符串如此，其他对象也是一样的。我们定义一个 A 类，将类的实例化对象存入集合，如代码 7-9 所示。

代码 7-9

```java
public class Test {
```

```java
public static void main(String[] args) {
    LinkedHashSet set = new LinkedHashSet();
    set.add(new A(1));
    set.add(new A(1));
    Iterator iterator = set.iterator();
    while(iterator.hasNext()) {
        System.out.print(iterator.next()+",");
    }
}
}

class A{
    private int num;
    public A(int num) {
        this.num = num;
    }
    @Override
    public String toString() {
        return "A [num=" + num + "]";
    }
}
```

运行结果如图 7-21 所示。

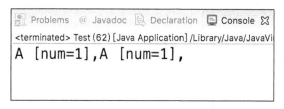

图 7-21

你可以看到两个 A 对象都保存到了集合中，也就是说当前集合不认为这两个对象相等。那么程序是如何来鉴别两个对象是否相等的呢？通过继承自 Object 类的 equals()方法来判断，Object 类的 equals()方法定义如图 7-22 所示。

```
public boolean equals(Object obj) {
    return (this == obj);
}
```

图 7-22

"=="表示比较两个对象的内存地址，所以虽然两个 A 对象的 num 值相等，也就是从内容的角度来看是相等的，但是内存地址不同，所以程序会认为是不相等的两个对象。现在有这样一个需求：只要两个对象的 num 值相等，就认为是同一个对象，那如何修改代码呢？首先需要说明一下用 LinkedHashSet 判断两个对象是否相等的原理，首先会判断两个对象的 hashCode 是否相等，什么是 hashCode？简单来说就是将对象的内部信息（如内存地址、属性值等），通过某种特定规则转换成一个散列值，也就是该对象的 hashCode。两个不同对象的 hashCode 可能相等，但是 hashCode 不相等的两个对象一定不是同一个对象。

所以集合在判断两个对象是否相等时，会先比较它们的 hashCode，如果不相等，则认为不是同一个对象，可以添加。如果 hashCode 相等，还不能认为两个对象就是相等的，需要通过 equals()方法进一步判断。如果 equals()方法为 true，则不会重复添加；如果 equals()方法为 false，则正常添加。先判断 hashCode 是否相等可以减少 equals()方法的调用，提高效率。所以两个 A 相等的前提是 hashCode 相等，且 equals()方法返回 true，修改后的程序如代码 7-10 所示。

代码 7-10

```
class A{
   ......
   @Override
   public boolean equals(Object obj) {
      // TODO Auto-generated method stub
      return true;
   }
   @Override
   public int hashCode() {
      // TODO Auto-generated method stub
      return 1;
   }
}
```

再次运行程序，结果如图 7-23 所示。

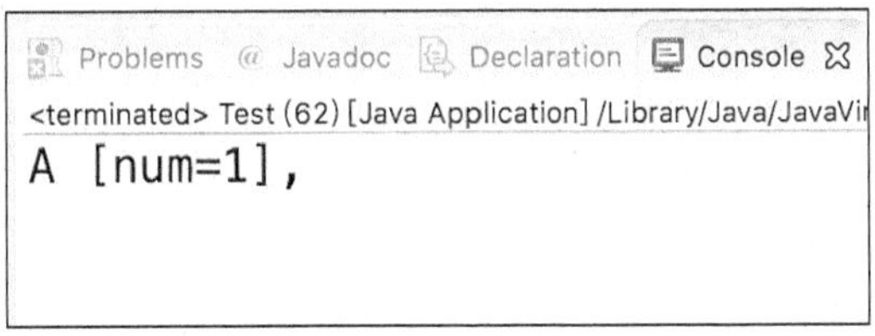
Problems
Javadoc
Declaration
Console
<terminated> Test (62) [Java Application] /Library/Java/JavaVi
A [num=1],

图 7-23

程序显示只存储了一个 A 对象，在 Set 的子接口中，除了 LinkedHashSet 可以存放有序元素之外，TreeSet 中保存的元素也是有序的，并且 TreeSet 的有序和 LinkedHashSet 的有序有所不同。LinkedHashSet 的有序是指元素的存储顺序和遍历顺序是一致的，即元素按什么顺序存进去，遍历时就按什么顺序输出。TreeSet 的有序是指集合内部会自动给所有的元素按照升序进行排列，即无论存入元素的顺序是什么，遍历时会按照升序进行输出。TreeSet 中的元素也是唯一的，定义如图 7-24 所示。

```
public class TreeSet<E> extends AbstractSet<E>
    implements NavigableSet<E>, Cloneable, java.io.Serializable
```

图 7-24

TreeSet 的使用如代码 7-11 所示。

代码 7-11

```
public class Test {
   public static void main(String[] args) {
      TreeSet treeSet = new TreeSet();
```

```
        treeSet.add(1);
        treeSet.add(3);
        treeSet.add(6);
        treeSet.add(2);
        treeSet.add(5);
        treeSet.add(4);
        treeSet.add(1);
        System.out.println("treeSet 的长度："+treeSet.size());
        System.out.println("遍历 treeSet");
        Iterator iterator = treeSet.iterator();
        while(iterator.hasNext()) {
            System.out.print(iterator.next()+",");
        }
        treeSet.remove(5);
        System.out.println("删除之后遍历 treeSet");
        iterator = treeSet.iterator();
        while(iterator.hasNext()) {
            System.out.print(iterator.next()+",");
        }
    }
}
```

运行结果如图 7-25 所示。

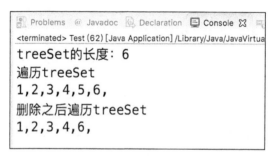

图 7-25

因为 TreeSet 内部会自动按照升序对元素进行排列，所以添加到 TreeSet 集合中的元素必须具备排序的功能，现在我们创建一个类 A，同时将 A 的实例化对象保存到 TreeSet 中，如代码 7-12 所示。

代码 7-12

```
public class Test {
    public static void main(String[] args) {
        TreeSet treeSet = new TreeSet();
        treeSet.add(new A(1));
        treeSet.add(new A(3));
        treeSet.add(new A(6));
        treeSet.add(new A(2));
        treeSet.add(new A(5));
        treeSet.add(new A(4));
        treeSet.add(new A(1));
        System.out.println("treeSet 的长度："+treeSet.size());
        System.out.println("遍历 treeSet");
        Iterator iterator = treeSet.iterator();
```

```java
        while(iterator.hasNext()) {
            System.out.print(iterator.next()+",");
        }
        treeSet.remove(new A(5));
        System.out.println("删除之后遍历 treeSet");
        iterator = treeSet.iterator();
        while(iterator.hasNext()) {
            System.out.print(iterator.next()+",");
        }
    }
}

class A{
    private int num;
    public A(int num) {
        this.num = num;
    }
}
```

运行结果如图 7-26 所示。

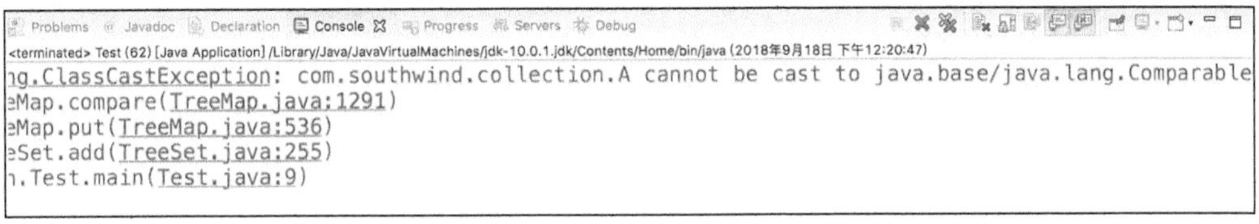

图 7-26

报错原因是 A 不具备排序的功能，如何解决呢？让 A 实现 Comparable 接口即可，如代码 7-13 所示。

代码 7-13

```java
class A implements Comparable{
    ……
    @Override
    public int compareTo(Object o) {
        // TODO Auto-generated method stub
        /**
         * A.compareTo(B)
         * 返回值：
         * 1 表示 A 大于 B
         * 0 表示 A 等于 B
         * -1 表示 A 小于 B
         */
        A a = (A) o;
        if(this.num > a.num) {
            return 1;
        }else if(this.num == a.num) {
            return 0;
        }else {
            return -1;
        }
    }
```

```
    @Override
    public String toString() {
        return "A [num=" + num + "]";
    }
}
```

再次运行，结果如图 7-27 所示。

```
Problems  Javadoc  Declaration  Console  Progress  Servers  Debug
<terminated> Test (62) [Java Application] /Library/Java/JavaVirtualMachines/jdk-10.0.1.jdk/Contents/Home/bin/java (2018年
treeSet的长度：6
遍历treeSet
A [num=1],A [num=2],A [num=3],A [num=4],A [num=5],A [num=6],
删除之后遍历treeSet
A [num=1],A [num=2],A [num=3],A [num=4],A [num=6],
```

图 7-27

7.5 Map 接口

7.5.1 Map 接口的定义

前面章节介绍的 Set，List 接口都是 Collection 的子接口。本章要介绍的 Map 接口是与 Collection 完全独立的另外一个体系。它们之间还有一个区别就是 Set、List、Collection 只能操作单个元素，而 Map 可以操作一对元素，因为 Map 中的元素都是以 key-value 的键值映射形式存储的，Map 接口的定义如图 7-28 所示。

```
public interface Map<K, V> {
    // Query Operations
```

图 7-28

Map 接口定义时使用了泛型，并且定义了两个泛型 K 和 V，K 表示 key，规定了键元素的数据类型，V 表示 value，规定了值元素的数据类型，Map 接口中的方法如表 7-4 所示。

表 7-4

方法	描述
int size()	获取集合长度
boolean isEmpty()	判断集合是否为空
boolean containsKey(Object key)	判断集合中是否存在某个 key 值
boolean containsValue(Object value)	判断集合中是否存在某个 value 值
V get(Object key)	取出集合中 key 对应的 value 值
V put(K key, V value)	向集合中存入一组 key-value 的元素
V remove(Object key)	删除集合中 key 对应的 value 值
void putAll(Map<? extends K, ? extends V> m)	向集合中添加另外一个 Map 集合
void clear()	清除集合中的所有元素

续表

方法	描述
Set<K> keySet()	取出集合中所有的 key，返回一个 Set 集合
Collection<V> values()	取出集合中所有的 value，返回一个 Collection 集合
Set<Map.Entry<K, V>> entrySet()	将 Map 对象转换为 Set 对象
int hashCode()	获取集合的散列值
boolean equals(Object o)	比较两个集合是否相等

7.5.2 Map 接口的实现类

Map 是一个接口，在实际开发中需要使用 Map 必须通过其实现类来完成实例化操作，Map 接口常用的实现类如下所示。

- HashMap：存储一组无序，key 不可重复，但 value 可重复的元素。
- Hashtable：存储一组无序，key 不可重复，但 value 可重复的元素。
- TreeMap：存储一组有序，key 不可重复，但 value 可重复的元素，可以按照 key 来排序。

HashMap 是 Map 接口的一个常用实现类，定义如图 7-29 所示。

```
public class HashMap<K,V> extends AbstractMap<K,V>
    implements Map<K,V>, Cloneable, Serializable {
```

图 7-29

HashMap 的使用如代码 7-14 所示。

代码 7-14

```java
public class Test {
    public static void main(String[] args) {
        HashMap<String,String> hashMap = new HashMap<String,String> ();
        hashMap.put("h", "Hello");
        hashMap.put("w", "World");
        hashMap.put("j", "Java");
        hashMap.put("s", "JavaSE");
        hashMap.put("m", "JavaME");
        hashMap.put("e", "JavaEE");
        System.out.println(hashMap);
        hashMap.remove("e");
        System.out.println("删除之后："+hashMap);
        hashMap.put("m", "Model");
        System.out.println("添加之后："+hashMap);
        if(hashMap.containsKey("a")) {
            System.out.println("集合中存在值为 a 的 key");
        }else {
            System.out.println("集合中不存在值为 a 的 key");
        }
        if(hashMap.containsValue("Java")) {
            System.out.println("集合中存在值为 Java 的 value");
```

```
        }else {
            System.out.println("集合中不存在值为 Java 的 value");
        }
        Set keys = hashMap.keySet();
        Iterator keysIterator = keys.iterator();
        System.out.print("集合中的 key: ");
        while(keysIterator.hasNext()) {
            String key = (String) keysIterator.next();
            System.out.print(key+",");
        }
        Collection values = hashMap.values();
        Iterator valuesIterator = values.iterator();
        System.out.print("集合中的 value: ");
        while(valuesIterator.hasNext()) {
            String value = (String) valuesIterator.next();
            System.out.print(value+",");
        }
        System.out.print("key-value: ");
        keysIterator = keys.iterator();
        while(keysIterator.hasNext()) {
            String key = (String) keysIterator.next();
            String value = hashMap.get(key);
            System.out.print(key+"-"+value+",");
        }
    }
}
```

运行结果如图 7-30 所示。

```
Problems  @ Javadoc  Declaration  Console ⊠  Progress  Servers  Debug
<terminated> Test (62) [Java Application] /Library/Java/JavaVirtualMachines/jdk-10.0.1.jdk/Contents/Home/bin/java (2018
{s=JavaSE, e=JavaEE, w=World, h=Hello, j=Java, m=JavaME}
删除之后: {s=JavaSE, w=World, h=Hello, j=Java, m=JavaME}
添加之后: {s=JavaSE, w=World, h=Hello, j=Java, m=Model}
集合中不存在值为a的key
集合中存在值为Java的value
集合中的key: s,w,h,j,m,
集合中的value: JavaSE,World,Hello,Java,Model,
key-value: s-JavaSE,w-World,h-Hello,j-Java,m-Model,
```

图 7-30

在 Map 的实现类中，用法与 HashMap 基本一样的实现类是 Hashtable。Hashtable 是较早推出的一个实现类，与 HashMap 的区别是：Hashtable 是线程安全的，但是性能较低。HashMap 是非线程安全的，但是性能较高，从实际开发角度来讲，HashMap 使用的频率更高，Hashtable 的定义如图 7-31 所示。

```
public class Hashtable<K,V>
    extends Dictionary<K,V>
    implements Map<K,V>, Cloneable, java.io.Serializable {
```

图 7-31

Hashtable 的使用如代码 7-15 所示。

代码 7-15

```java
public class Test {
    public static void main(String[] args) {
        Hashtable<String,String> hashtable = new Hashtable<String,String> ();
        hashtable.put("h", "Hello");
        hashtable.put("w", "World");
        hashtable.put("j", "Java");
        hashtable.put("s", "JavaSE");
        hashtable.put("m", "JavaME");
        hashtable.put("e", "JavaEE");
        System.out.println(hashtable);
        hashtable.remove("e");
        System.out.println("删除之后："+hashtable);
        hashtable.put("m", "Model");
        System.out.println("添加之后："+hashtable);
        if(hashtable.containsKey("a")) {
            System.out.println("集合中存在值为a 的key");
        }else {
            System.out.println("集合中不存在值为a 的key");
        }
        if(hashtable.containsValue("Java")) {
            System.out.println("集合中存在值为Java 的value");
        }else {
            System.out.println("集合中不存在值为Java 的value");
        }
        Set keys = hashtable.keySet();
        Iterator keysIterator = keys.iterator();
        System.out.print("集合中的key：");
        while(keysIterator.hasNext()) {
            String key = (String) keysIterator.next();
            System.out.print(key+",");
        }
        Collection values = hashtable.values();
        Iterator valuesIterator = values.iterator();
        System.out.print("集合中的value：");
        while(valuesIterator.hasNext()) {
            String value = (String) valuesIterator.next();
            System.out.print(value+",");
        }
        System.out.print("key-value：");
        keysIterator = keys.iterator();
        while(keysIterator.hasNext()) {
            String key = (String) keysIterator.next();
            String value = hashtable.get(key);
            System.out.print(key+"-"+value+",");
        }
    }
}
```

运行结果如图 7-32 所示。

无论是 HashMap 还是 Hashtable，保存的数据都是无序的，我们可以从上面两个例子的输出结果看出这一点。Map 的另外一个实现类 TreeMap 主要功能就是按照 key 对集合中的数据进行排序，TreeMap 的定义如图 7-33 所示。

```
{m=JavaME, w=World, j=Java, s=JavaSE, h=Hello, e=JavaEE}
删除之后：{m=JavaME, w=World, j=Java, s=JavaSE, h=Hello}
添加之后：{m=Model, w=World, j=Java, s=JavaSE, h=Hello}
集合中不存在值为a的key
集合中存在值为Java的value
集合中的key: m,w,j,s,h,
集合中的value: Model,World,Java,JavaSE,Hello,
key-value: m-Model,w-World,j-Java,s-JavaSE,h-Hello,
```

图 7-32

```
public class TreeMap<K,V>
    extends AbstractMap<K,V>
    implements NavigableMap<K,V>, Cloneable, java.io.Serializable
```

图 7-33

TreeMap 的使用如代码 7-16 所示。

代码 7-16

```java
public class Test {
    public static void main(String[] args) {
        TreeMap<Integer,String> treeMap = new TreeMap<Integer,String>();
        treeMap.put(3, "Java");
        treeMap.put(5, "JavaME");
        treeMap.put(1, "Hello");
        treeMap.put(6, "JavaEE");
        treeMap.put(2, "World");
        treeMap.put(4, "JavaSE");
        Set keys = treeMap.keySet();
        Iterator keysIterator = keys.iterator();
        while(keysIterator.hasNext()) {
            Integer key = (Integer) keysIterator.next();
            String value = treeMap.get(key);
            System.out.print(key+"-"+value+",");
        }
    }
}
```

运行结果如图 7-34 所示。

```
1-Hello,2-World,3-Java,4-JavaSE,5-JavaME,6-JavaEE,
```

图 7-34

可以看到，无论向集合中保存数据时的顺序如何，TreeMap 内部会自动按照 key 升序对数据进行排序。此时的 key 是 Integer 类型的，那如果 key 不是一个可自动排序的数据类型，如自定义的 User 类型，TreeMap 能否对其进行排序呢？具体实现如代码 7-17 所示。

代码 7-17

```java
public class Test {
    public static void main(String[] args) {
        TreeMap<User,String> treeMap = new TreeMap<User,String>();
        treeMap.put(new User(3,"Java"), "Java");
        treeMap.put(new User(5,"JavaME"), "JavaME");
        treeMap.put(new User(1,"Hello"), "Hello");
        treeMap.put(new User(6,"JavaEE"), "JavaEE");
        treeMap.put(new User(2,"World"), "World");
        treeMap.put(new User(4,"JavaSE"), "JavaSE");
        Set keys = treeMap.keySet();
        Iterator keysIterator = keys.iterator();
        while(keysIterator.hasNext()) {
            User key = (User) keysIterator.next();
            String value = treeMap.get(key);
            System.out.println(key+"-"+value);
        }
        System.out.println("集合中第一个 Entry: "+treeMap.firstEntry());
        System.out.println("集合中第一个 key: "+treeMap.firstKey());
        System.out.println("集合中最后一个 Entry: "+treeMap.lastEntry());
        System.out.println("集合中最后一个 key: "+treeMap.lastKey());
        System.out.println("集合中比 new User(3,Java)大的最小 key 值: "+treeMap.higherKey(new User(3,"Java")));
        System.out.println("集合中比 new User(3,Java)小的最大 key 值:"+treeMap.lowerKey(new User(3,"Java")));
        System.out.println("集合中比 new User(3,Java)大的最小的 key-value 对:"+treeMap.higherEntry(new User(3,"Java")));
        System.out.println("集合中比 new User(3,Java)小的最大的 key-value 对:"+treeMap.lowerEntry(new User(3,"Java")));
        System.out.println("截取之后的集合: "+treeMap.subMap(new User(3,"Java"), new User(5,"JavaME")));
    }
}

class User{
    private int id;
    private String name;
    //getter、setter、有参构造函数、重写 toString
}
```

运行结果如图 7-35 所示。

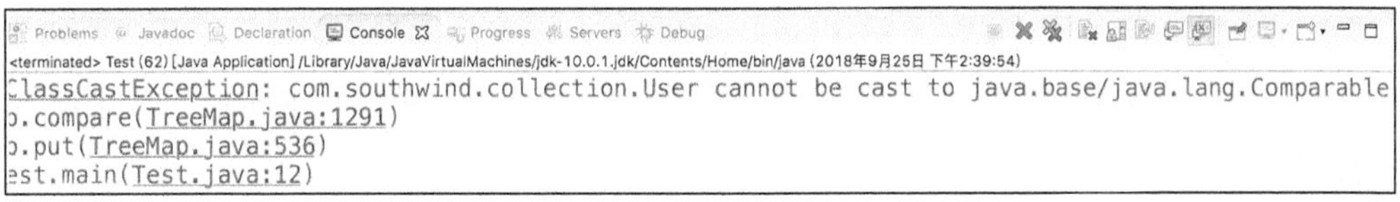

图 7-35

程序运行报错，错误原因是 User 的实例化对象无法进行排序，解决方法是让 User 类实现 Comparable 接口，并在 compareTo()方法中实现对象的排序规则，如代码 7-18 所示。

代码 7-18

```java
class User implements Comparable{
```

```java
......
@Override
public int compareTo(Object o) {
    // TODO Auto-generated method stub
    /**
     * A.compareTo(B)
     * 返回值：
     * 1 表示 A 大于 B
     * 0 表示 A 等于 B
     * -1 表示 A 小于 B
     */
    User user = (User) o;
    if(this.id > user.id) {
        return 1;
    }else if(this.id == user.id) {
        return 0;
    }else {
        return -1;
    }
}
```

再次运行程序，结果如图 7-36 所示。

```
<terminated> Test (62) [Java Application] /Library/Java/JavaVirtualMachines/jdk-10.0.1.jdk/Contents/Home/bin/java (2018年9月25日 下午4:08:36)
User [id=1, name=Hello]-Hello
User [id=2, name=World]-World
User [id=3, name=Java]-Java
User [id=4, name=JavaSE]-JavaSE
User [id=5, name=JavaME]-JavaME
User [id=6, name=JavaEE]-JavaEE
集合中第一个Entry: User [id=1, name=Hello]=Hello
集合中第一个key: User [id=1, name=Hello]
集合中最后一个Entry: User [id=6, name=JavaEE]=JavaEE
集合中最后一个key: User [id=6, name=JavaEE]
集合中比new User(3,"Java")大的最小key值: User [id=4, name=JavaSE]
集合中比new User(3,"Java")小的最大key值: User [id=2, name=World]
集合中比new User(3,"Java")大的最小的key-value对: User [id=4, name=JavaSE]=JavaSE
集合中比new User(3,"Java")小的最大的key-value对: User [id=2, name=World]=World
截取之后的集合: {User [id=3, name=Java]=Java, User [id=4, name=JavaSE]=JavaSE}
```

图 7-36

通过结果可以看到，此时集合以 User 对象的 id 值升序排列为规则，对集合内的元素进行了排序，准确地讲应该是对集合中的 key 值进行排序，并且 TreeMap 支持对集合数据的多种操作。

7.6 Collections 工具类

集合除了可以存储数据，也提供了很多方法来对数据进行操作，但是这些方法都有其局限性，实际操作起来不是很方便。JDK 为我们提供来一个工具类 Collections，专门用来操作集合，例如添加元素、对元素进行排序、替换元素等。Collections 和 Arrays 很类似，Arrays 是针对数组的工具类，Collections 是针对集合的工具类，Collections 的定义如图 7-37 所示。

```
public class Collections {
    // Suppresses default constructor,
```

图 7-37

Collections 的常用方法描述如表 7-5 所示。

表 7-5

方法	描述
public static \<T extends Comparable\<? super T>> void sort(List\<T> list)	根据集合泛型对应的类实现的 Comparable 接口对集合进行排序
public static \<T> void sort(List\<T> list, Comparator\<? super T> c)	根据 Comparator 接口对集合进行排序
public static \<T> int binarySearch(List\<? extends Comparable\<? super T>> list, T key)	查找元素在集合中的下标，要求集合元素必须是升序排列
private static \<T> T get(ListIterator\<? extends T> i, int index)	根据下标找到集合中的元素
public static void reverse(List\<?> list)	对集合元素的顺序进行反转
public static void swap(List\<?> list, int i, int j)	交换集合中指定位置的两个元素
public static \<T> void fill(List\<? super T> list, T obj)	将集合中的所有元素替换为 obj
public static \<T> T min(Collection\<? extends T> coll, Comparator\<? super T> comp)	根据 Comparator 接口对集合进行排序，返回集合中的最小值
public static \<T> T max(Collection\<? extends T> coll, Comparator\<? super T> comp)	根据 Comparator 接口对集合进行排序，返回集合中的最大值
public static \<T> boolean replaceAll(List\<T> list, T oldVal, T newVal)	将集合中的所有 oldVal 替换为 newVal
public static \<T> boolean addAll(Collection\<? super T> c, T... elements)	向集合添加元素

Collections 常用方法的使用如代码 7-19 所示。

代码 7-19

```java
public class Test {
    public static void main(String[] args) {
        ArrayList list = new ArrayList();
        list.add("Hello");
        list.add("World");
        System.out.println("添加之前的集合："+list);
        Collections.addAll(list, "Java","JavaSE","JavaME");
        System.out.println("添加之后的集合："+list);
        Collections.reverse(list);
        System.out.println("反转之后的集合："+list);
        Collections.swap(list, 1, 3);
        System.out.println("交换之后的集合："+list);
        Collections.sort(list);
        System.out.println("先对集合进行排序："+list);
        int index = Collections.binarySearch(list, "JavaME");
        System.out.println("JavaME 在集合中的下标："+index);
```

```
            Collections.replaceAll(list, "Java", "Collections");
            System.out.println("替换之后的集合:"+list);
            list = new ArrayList();
            Collections.addAll(list,new User(1,"张三",30),new User(2,"李四",26),new 
User(3,"王五",18));
            System.out.println("排序之前的集合:"+list);
            Collections.sort(list,new Comparator() {
                @Override
                public int compare(Object o1, Object o2) {
                    // TODO Auto-generated method stub
                    User user1 = (User) o1;
                    User user2 = (User) o2;
                    if(user1.getAge() > user2.getAge()) {
                        return 1;
                    }else if(user1.getAge() == user2.getAge()) {
                        return 0;
                    }else {
                        return -1;
                    }
                }
            });
            System.out.println("排序之后的集合:"+list);
    }
}

class User {
    ……
}
```

再次运行程序,结果如图 7-38 所示。

```
添加之前的集合:[Hello, World]
添加之后的集合:[Hello, World, Java, JavaSE, JavaME]
反转之后的集合:[JavaME, JavaSE, Java, World, Hello]
交换之后的集合:[JavaME, World, Java, JavaSE, Hello]
先对集合进行排序:[Hello, Java, JavaME, JavaSE, World]
JavaME在集合中的下标: 2
替换之后的集合:[Hello, Collections, JavaME, JavaSE, World]
排序之前的集合:[User [name=张三, age=30], User [name=李四, age=26], User [name=王五, age=18]]
排序之后的集合:[User [name=王五, age=18], User [name=李四, age=26], User [name=张三, age=30]]
```

图 7-38

7.7 泛型

7.7.1 泛型的概念

泛型(Generics)是指在类定义时不指定类中信息的具体数据类型,而是用一个标识符来代替,当外部实例化对象时来指定具体的数据类型。有了泛型,我们就可以在定义类或者接口时不明确指定类中信息的具体数据类型,在实例化时再来指定具体的数据类型。这样极大地提高了类的扩展性,一个类可以装载各种不同的数据类型,泛型可以指代类中的成员变量数据类型，方法的返回值数据类型以及方法的参数数据类型。

为什么要使用泛型呢？来看下面这个例子,我们知道一个集合可以存储不同数据类型的数据,实例化一个 ArrayList 集合对象,并向该集合中保存两个数据,int 类型的 1 和 String 类型的"Hello"，如代码 7-20 所示。

代码 7-20

```
public class Test {
    public static void main(String[] args) {
        ArrayList list = new ArrayList();
        list.add(1);
        list.add("Hello");
        for(int i = 0; i < list.size(); i++) {
            int num = (int) list.get(i);
            System.out.println(num+1);
        }
    }
}
```

运行结果如图 7-39 所示。

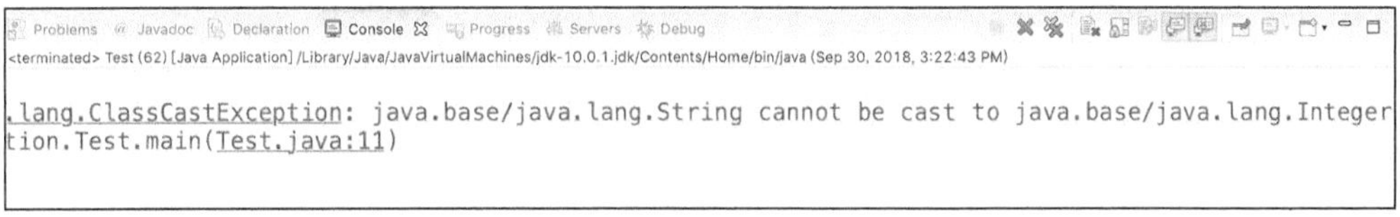

图 7-39

报错原因在于无法将 String 类型的"Hello"转为 int 类型，这就是没有设置泛型所带来的数据不安全问题。因为 list 对数据类型没有要求，任意数据类型都能存入，集合内部是以 Object 类型来保存各种数据的，也是多态的一种体现。存入没有问题，但是在取数据的时候就出问题了，因为不同的数据类型之间往往不能进行类型转换，除非两个数据有继承关系，或者具有接口实现的关系，否则强制进行转换就会抛出异常。

使用泛型就可以避免数据不安全的隐患，接口是支持泛型的。所以我们在实例化 ArrayList 对象的时候就指定泛型为 Integer，这样就限制了可以存入集合的数据，除了 Integer 类型以外的数据类型无法存入集合，当然指定数据类型的子类是可以存入的，这样就保证了集合中数据类型的统一性。在取数据的时候不会抛出数据转换类型失败的异常，同时在指定泛型后，集合内部就会以指定的数据类型来保存所有数据，取数据时也就省去了强制类型转换的步骤，修改如代码 7-21 所示。

代码 7-21

```
public class Test {
    public static void main(String[] args) {
        ArrayList<Integer> list = new ArrayList<Integer>();
        list.add(1);
        list.add("Hello");
        for(int i = 0; i < list.size(); i++) {
            int num = (int) list.get(i);
```

```
        System.out.println(num+1);
    }
  }
}
```

代码出现编译错误，如图 7-40 所示。

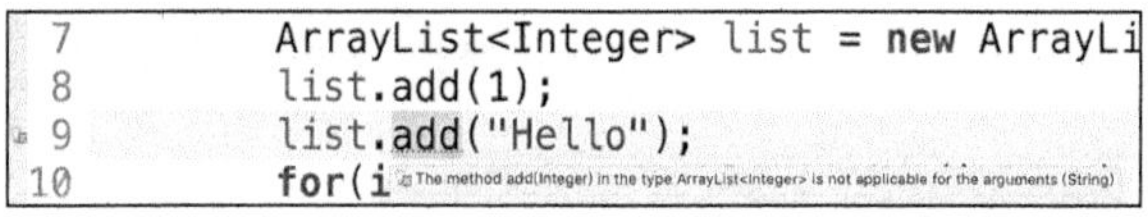

图 7-40

因为此时 list 指定了泛型 Integer，所以 String 类型的"Hello"无法存入集合，修改如代码 7-22 所示。

代码 7-22

```
public class Test {
    public static void main(String[] args) {
        ArrayList<Integer> list = new ArrayList<Integer>();
        list.add(1);
        list.add(2);
        for(int i = 0; i < list.size(); i++) {
            int num = (int) list.get(i);
            System.out.println(num+1);
        }
    }
}
```

运行结果如图 7-41 所示。

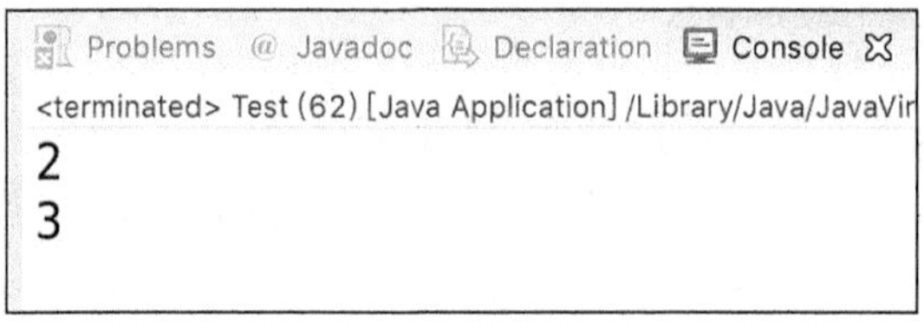

图 7-41

7.7.2 泛型的应用

我们除了可以在实例化集合时指定泛型外，自定义的类也可以添加泛型，基本语法如下：

```
访问权限修饰符 class 类名<泛型标识1，泛型标识2...>{
    访问权限修饰符 泛型标识 属性名；
    访问权限修饰符 泛型标识 方法名(泛型标识 参数名...){}
}
```

例如我们自定义一个表示时间的类 Time，如代码 7-23 所示。

代码 7-23

```java
public class Time<T> {
    private T value;
    public T getValue() {
        return value;
    }
    public void setValue(T value) {
        this.value = value;
    }
}

public class Test {
    public static void main(String[] args) {
        Time<Integer> time1 = new Time<Integer>();
        time1.setValue(10);
        System.out.println("现在的时间是："+time1.getValue());
        Time<String> time2 = new Time<String>();
        time2.setValue("十点整");
        System.out.println("现在的时间是："+time2.getValue());
    }
}
```

运行结果如图 7-42 所示。

图 7-42

在定义一个类时可以同时指定多个泛型标识，如代码 7-24 所示。

代码 7-24

```java
public class Time<H,M,S> {
    private H hour;
    private M minute;
    private S second;
    //getter、setter 方法
}

public class Test {
    public static void main(String[] args) {
        Time<String,Integer,Float> time = new Time<String,Integer,Float>();
        time.setHour("十点");
        time.setMinute(10);
        time.setSecond(10.0f);
        System.out.println("现在的时间是："+time.getHour()+":"+time.getMinute()+":"+time.getSecond());
    }
```

        }

运行结果如图 7-43 所示。

图 7-43

7.7.3 泛型通配符

如果我们在定义一个参数为 ArrayList 类型的方法时，希望该方法既可以接收泛型为 String 的集合参数，也可以接收泛型为 Integer 的集合参数，那应该怎么处理呢？可以用多态的思想来定义该方法，即使用 Object 来定义参数泛型，如代码 7-25 所示。

代码 7-25

```
public class Test {
   public static void main(String[] args) {
      ArrayList<String> list1 = new ArrayList<String>();
      ArrayList<Integer> list2 = new ArrayList<Integer>();
      test(list1);
      test(list2);
   }
   public static void test(ArrayList<Object> list) {
      System.out.println(list);
   }
}
```

此时，代码出现了编译错误，如图 7-44 所示。

```
 9        test(list1);
10        test(list2);
11     }
```

图 7-44

list1 和 list2 无法匹配 test() 方法的参数类型，String 和 Integer 在泛型引用中不能转换为 Object。所以 test() 方法的参数泛型不能设置为 Object，如何解决呢？这个问题可以通过泛型通配符来处理，用?表示当前未知的泛型类型，如代码 7-26 所示。

代码 7-26

```
public class Test {
   public static void main(String[] args) {
      ArrayList<String> list1 = new ArrayList<String>();
      ArrayList<Integer> list2 = new ArrayList<Integer>();
      test(list1);
      test(list2);
   }
```

```java
    public static void test(ArrayList<?> list) {
        System.out.println(list);
    }
}
```

ArrayList<?>表示可以使用任意的泛型类型对象,这样 test()方法就具有通用性了。

7.7.4 泛型上限和下限

我们在使用泛型时,往往数据类型会有限制,只能使用一种具体的数据类型。如果希望在此基础之上进行适量扩容,可以通过泛型上限和下限来完成。泛型上限表示实例化时的具体数据类型,可以是上限类型的子类或者是上限类型本身,用 extends 关键字来修饰。泛型下限表示实例化时的具体数据类型可以是下限类型的父类或者是下限类型本身,用 super 关键字来修饰,基本语法如下。

泛型上限:类名<泛型标识 extends 上限类名>

泛型下限:类名<泛型标识 super 下限类名>

具体实现如代码 7-27 所示。

代码 7-27

```java
public class Time<T> {
    public static void main(String[] args) {
        test(new Time<Integer>());
        test(new Time<String>());
        test2(new Time<String>());
        test2(new Time<Integer>());
    }
    /*
     * 参数的泛型只能是 Number 或者其子类,即 Number, Byte, Short
     * Long, Integer, Float, Double
     */
    public static void test(Time<? extends Number> time) {

    }
    /*
     * 参数的泛型只能是 String 或者其父类,即 String 和 Object
     */
    public static void test2(Time<? super String> time) {

    }
}
```

代码出现编译错误,如图 7-45 所示。

```
public static void main(String[] args) {
    test(new Time<Integer>());
    test(new Time<String>());        ← String 不是 Number 的子类
    test2(new Time<String>());
    test2(new Time<Integer>());      ← Integer 不是 String 的父类
}
```

图 7-45

7.7.5 泛型接口

我们在定义类时可以添加泛型，在定义接口时也可以添加泛型。声明泛型接口的语法和声明泛型类很相似，在接口名后加上<T>即可，基本语法：访问权限修饰符 interface 接口名<泛型标识>，具体实现如代码 7-28 所示。

代码 7-28

```java
public interface MyInterface<T> {
    public T getValue();
}
```

实现泛型接口有两种方式，一种是实现类在定义时继续使用泛型标识，另一种是实现类在定义时直接给出具体的数据类型，实现如代码 7-29 所示。

代码 7-29

```java
public class MyInterfaceImpl<T> implements MyInterface<T>{
    private T obj;
    //getter、setter方法

    public MyInterfaceImpl(T obj) {
        this.obj = obj;
    }
    @Override
    public T getValue() {
        // TODO Auto-generated method stub
        return this.obj;
    }
}


public class MyInterfaceImpl2 implements MyInterface<String>{
    private String obj;
    //getter、setter方法

    public MyInterfaceImpl2(String obj) {
        this.obj = obj;
    }
    @Override
    public String getValue() {
        // TODO Auto-generated method stub
        return this.obj;
    }
}
```

两种不同实现类的实例化方式也不同，一种需要在实例化时指定具体的数据类型，另一种在实例化时不需要指定具体的数据类型，如代码 7-30 所示。

代码 7-30

```java
public class Test {
    public static void main(String[] args) {
```

```java
        MyInterface<String> myInterface = new MyInterfaceImpl<String>("接口");
        String value1 = myInterface.getValue();
        MyInterface myInterface2 = new MyInterfaceImpl2("接口");
        String value2 = myInterface2.getValue();
    }
}
```

7.8 综合练习

对本书第 3 章的综合练习进行代码重构,要求使用面向对象编程思想,同时使用集合替代数组来存储数据,具体实现如代码 7-31 所示。

代码 7-31

```java
public class User{
    private String name;
    private int age;
    private String state;
    //getter、setter 方法

    public User(String name, int age, String state) {
        super();
        this.name = name;
        this.age = age;
        this.state = state;
    }
}


public class Test {
    //初始化用户集合
    private static List<User> userList;
    static {
        userList = new ArrayList<User>();
        userList.add(new User("张三",22,"正常"));
        userList.add(new User("李四",23,"正常"));
        userList.add(new User("王五",20,"正常"));
        userList.add(new User("小明",22,"正常"));
    }
    public static void main(String[] args) {
        Scanner scanner = new Scanner(System.in);
        int num;
        do{
            System.out.println("欢迎使用用户管理系统");
            System.out.println("1.查询用户");
            System.out.println("2.添加用户");
            System.out.println("3.删除用户");
            System.out.println("4.账号冻结");
            System.out.println("5.账号解封");
            System.out.println("6.退出系统");
            System.out.print("请选择：");
            num = scanner.nextInt();
```

```java
            switch(num){
                case 1:
                    System.out.println("------查询用户------");
                    System.out.println("编号\t\t名称\t\t年龄\t\t状态");
                    for(int i = 0; i < userList.size(); i++) {
                        User user = userList.get(i);
                        System.out.println((i+1)+"\t\t"+user.getName()+"\t\t"+user.getAge()+"\t\t"+user.getState());
                    }
                    System.out.print("输入0返回：");
                    num = scanner.nextInt();
                    break;
                case 2:
                    System.out.println("------添加用户------");
                    System.out.print("请输入用户名称：");
                    String name = scanner.next();
                    boolean flag = false;
                    //判断该用户是否存在
                    for(int i = 0; i < userList.size(); i++){
                        User user = userList.get(i);
                        if(user.getName().equals(name)){
                            System.out.println("该用户已存在！");
                            flag = true;
                            break;
                        }
                    }
                    //添加用户
                    if(!flag){
                        System.out.print("请输入用户年龄：");
                        int age = scanner.nextInt();
                        userList.add(new User(name,age,"正常"));
                        System.out.println(name+"添加成功!");
                    }
                    System.out.print("输入0返回：");
                    num = scanner.nextInt();
                    break;
                case 3:
                    System.out.println("------删除用户------");
                    System.out.print("请输入用户名称：");
                    name = scanner.next();
                    //判断该用户是否存在
                    boolean flag2 = false;
                    for(int i = 0; i < userList.size(); i++){
                        User user = userList.get(i);
                        if(user.getName().equals(name)) {
                            flag2 = true;
                            userList.remove(user);
                        }
                    }
                    if(!flag2){
                        System.out.println(name+"不存在，请重新输入！");
                    }else{
                        System.out.println(name+"删除成功!");
                    }
                    System.out.print("输入0返回：");
                    num = scanner.nextInt();
                    break;
```

```java
                case 4:
                    System.out.println("------账号冻结------");
                    System.out.print("请输入用户名称：");
                    name = scanner.next();
                    //判断该用户是否存在
                    boolean flag3 = false;
                    for(int i = 0; i < userList.size(); i++){
                        User user = userList.get(i);
                        if(user.getName().equals(name)) {
                            flag3 = true;
                            if(user.getState().equals("冻结")){
                                System.out.println(name+"已冻结!");
                            }else{
                                user.setState("冻结");
                                System.out.println(name+"冻结成功!");
                            }
                            break;
                        }
                    }
                    if(!flag3){
                        System.out.println(name+"不存在，请重新输入!");
                    }
                    System.out.print("输入 0 返回：");
                    num = scanner.nextInt();
                    break;
                case 5:
                    System.out.println("------账号解封------");
                    System.out.print("请输入用户名称：");
                    name = scanner.next();
                    //判断该用户是否存在
                    boolean flag4 = false;
                    for(int i = 0; i < userList.size(); i++){
                        User user = userList.get(i);
                        if(user.getName().equals(name)) {
                            flag4 = true;
                            if(user.getState().equals("正常")){
                                System.out.println(name+"状态正常!");
                            }else{
                                user.setState("正常");
                                System.out.println(name+"解封成功!");
                            }
                            break;
                        }
                    }
                    if(!flag4){
                        System.out.println(name+"不存在，请重新输入!");
                    }
                    System.out.print("输入 0 返回：");
                    num = scanner.nextInt();
                    break;
                case 6:
                    System.out.println("感谢使用用户管理系统!");
                    return;
            }
        }while(num == 0);
    }
}
```

7.9 小结

本章为大家系统讲解了 Java 中的集合框架，其中包含了很多种不同的分类，大体可分为两类：Collection 和 Map。Collection 用于保存单列数据，又可分为 List 和 Set。Map 通过键值对映射的形式保存双列数据。从数据结构上来看 Collection 和 Map 是完全不同的两种集合，并且它们都是接口，我们在实际开发中需要使用具体的实现类来完成相关操作。常用的实现类包括 ArrayList、LinkedList、HashSet、TreeSet、HashMap、Hashtable 等，这一章的知识在实际开发中使用频率较高，业务层面的数据存储，乃至模块之间的数据传递都是通过集合来实现的。

第 8 章　实用类

> Java 语言为开发者提供了很多操作类库,在实际开发中使用这些类库可以让编程变得更加方便简单。如我们之前讲的 Arrays、Collections、String 类等,都是 Java 提供给开发者的类库。本章将为大家详细讲解实际开发中常用的类库。

8.1　枚举

枚举(Enum)是一种有确定取值区间的数据类型,它本质上是一种类,具有简洁、安全、方便等特点。可以这样理解,枚举的值被约束到一个特定的范围,只能取这个范围以内的值。

在了解枚举的具体概念之前,我们先来思考一个问题:为什么要有枚举呢?我们在描述对象的一些属性特征时,可选择的值是有一个特定范围的,即该属性的值不能随便定义。比如性别就只能选择男和女;一周只有周一到周日这 7 个选择,不可能出现星期八;一年只有春、夏、秋、冬 4 个季节等。出于对数据的安全性考虑,类似这种有特定取值范围的数据我们就可以使用枚举来描述,这就是使用枚举的意义所在。枚举指由一组常量组成的类型,指定了一个取值区间,我们只能从这个区间中取值。再来看枚举的定义,在没有枚举之前,我们需要定义一个类来描述周一到周日,可以通过定义静态常量的方式来完成,如代码 8-1 所示。

代码 8-1

```java
public class Week {
    public static final int MONDAY = 0;
    public static final int TUESDAY = 1;
    public static final int WEDNESDAY = 2;
    public static final int THURSDAY = 3;
    public static final int FRIDAY = 4;
    public static final int SATURDAY = 5;
    public static final int SUNDAY = 6;
}
```

这种方式是可以完成需求的,但是编写起来会比较麻烦,用 int 类型的数据来描述周几也不是很直观,如果使用枚举类型就会方便很多,如代码 8-2 所示。

代码 8-2

```
public enum Week {
    MONDAY,TUESDAY,WEDNESDAY,THURSDAY,FRIDAY,SATURDAY,SUNDAY;
}
```

枚举的定义与类很相似，使用 enum 关键字来描述，基本语法如下：

```
public enum 枚举名{
值1,值2,值3...
}
```

需要注意的是枚举中的常量使用逗号进行分割，看到这里有的读者可能会有疑惑，枚举中的常量值是什么呢？枚举中的每一个常量都对应的是一个枚举实例，只不过表示的含义不同。拿上面这个例子来说，Java 在编译期会帮我们生成一个 Week 类，并且继承自 java.lang.Enum，被 final 修饰，表示该类不可被继承。同时还生成了 7 个 Week 的实例对象分别对应枚举中定义的 7 个日期，因为枚举的静态常量直接对应其实例化对象，所以对于枚举的使用如代码 8-3 所示。

代码 8-3

```
public class Test {
    public static void main(String[] args) {
        Week week = Week.MONDAY;
        System.out.println(week);
    }
}
```

运行结果如图 8-1 所示。

图 8-1

编译期生成的类如代码 8-4 所示。

代码 8-4

```
final class Week extends Enum {
    public static final Week MONDAY;
    public static final Week TUESDAY;
    public static final Week WEDNESDAY;
    public static final Week THURSDAY;
    public static final Week FRIDAY;
    public static final Week SATURDAY;
    public static final Week SUNDAY;
    private static final Week $VALUES[];
static {
    MONDAY = new Week("MONDAY", 0);
    TUESDAY = new Week("TUESDAY", 1);
    WEDNESDAY = new Week("WEDNESDAY", 2);
```

```
        THURSDAY = new Week("THURSDAY", 3);
        FRIDAY = new Week("FRIDAY", 4);
        SATURDAY = new Week("SATURDAY", 5);
        SUNDAY = new Week("SUNDAY", 6);
        $VALUES = (new Week[] { MONDAY, TUESDAY, WEDNESDAY, THURSDAY, FRIDAY, SATURDAY, SUNDAY });
    }
    public static Week[] values() {
        return (Week[])$VALUES.clone();
    }
    public static Week valueOf(String s) {
        return (Week)Enum.valueOf(com/southwind/Week, s);
    }
    private Week(String s, int i) {
        super(s, i);
    }
}
```

我们来解读一下这个类,首先定义了 7 个 Week 类型的静态常量和一个 Week 类型的静态数组常量。同时定义了一个私有的构造函数,String 类型的参数即当前枚举对象的值,int 类型的参数为它的下标。静态代码块中通过私有构造函数对 7 个静态常量以及静态数组常量赋值,所以代码 8-4 中打印的枚举值其实就是创建该对象时传入的 String 类型参数,如"MONDAY"。同时该类还为我们提供了两个静态方法 values() 和 valueOf(String s),values() 方法可以返回该枚举的所有常量,vauleOf(String s)可以通过字符串 s 创建对应的枚举对象,具体操作如代码 8-5 所示。

代码 8-5

```java
public class Test {
    public static void main(String[] args) {
        Week[] weeks = Week.values();
        for(Week week:weeks) {
            System.out.println(week);
        }
        System.out.println("----------");
        Week week = Week.valueOf("MONDAY");
        System.out.println(week);
    }
}
```

运行结果如图 8-2 所示。

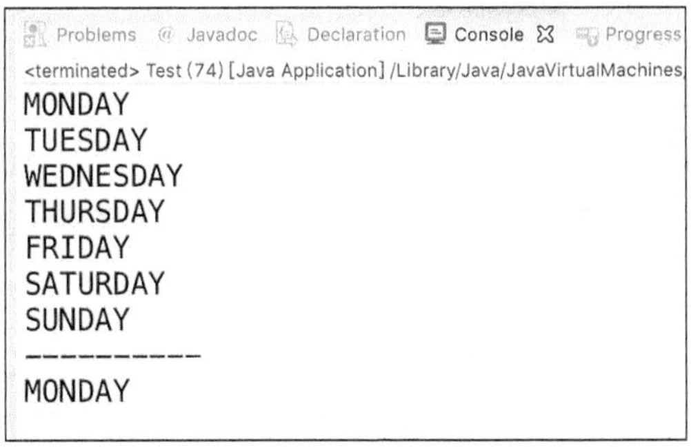

图 8-2

8.2 Math

Math 类为开发者提供了一系列的数学方法，同时还提供了两个静态常量 E（自然对数的底数）和 PI（圆周率），以满足项目研发中对于数学运算的要求，Math 类中的所有方法全部都是静态的，通过类名直接调用。Math 类比较简单，我们直接来看 Math 方法的使用，如代码 8-6 所示。

代码 8-6

```java
public class Test {
    public static void main(String[] args) {
        System.out.println("常量E："+Math.E);
        System.out.println("常量PI："+Math.PI);
        System.out.println("9 的平方根："+Math.sqrt(9));
        System.out.println("8 的立方根："+Math.cbrt(8));
        System.out.println("2 的 3 次平方："+Math.pow(2,3));
        System.out.println("较大值："+Math.max(6.3,3.5));
        System.out.println("较小值："+Math.min(6.3,3.5));
        System.out.println("-10.3 的绝对值："+Math.abs(-10.3));
        System.out.println("ceil(10.001)："+Math.ceil(10.001));
        System.out.println("floor(10.999): "+Math.floor(10.999));
        System.out.println("随机数："+Math.random());
        System.out.println("5.6 四舍五入："+Math.rint(5.6));
        System.out.println("5.6f 四舍五入："+Math.round(5.6f));
        System.out.println("5.6 四舍五入："+Math.round(5.6));
    }
}
```

运行结果如图 8-3 所示。

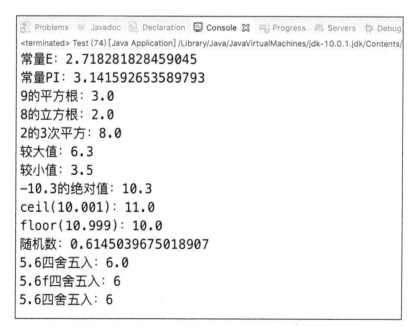

图 8-3

8.3 Random

Random 是用来产生一个随机数的类,并且可以任意指定一个区间,在此区间内产生一个随机数。Random 类的常用方法如表 8-1 所示。

表 8-1

方法	描述
public Random()	创建一个无参的随机数构造器,使用系统时间(ms)作为默认种子
public Random(long seed)	使用 long 数据类型的种子创建一个随机数构造器
public boolean nextBoolean()	返回下一个伪随机数,它取自此随机数生成器序列的均匀分布的 boolean 值
public double nextDouble()	返回下一个伪随机数,它取自此随机数生成器序列的,在 0.0 和 1.0 之间均匀分布的 double 值
public float nextFloat()	返回下一个伪随机数,它取自此随机数生成器序列的,在 0.0 和 1.0 之间均匀分布的 float 值
public int nextInt()	返回下一个伪随机数,它取自此随机数生成器的序列中均匀分布的 int 值
public int nextInt(int n)	返回一个伪随机数,它取自此随机数生成器序列的,在 0 和 n 之间均匀分布的 int 值
public long nextLong()	返回下一个伪随机数,它取自此随机数生成器序列的均匀分布的 long 值
public synchronized void setSeed(long seed)	使用单个 long 种子设置此随机数生成器的种子

Random 类的具体使用如代码 8-7 所示。

代码 8-7

```
public class Test {
   public static void main(String[] args) {
      Random random = new Random();
      for (int i = 1; i <= 3; i++) {
         boolean flag = random.nextBoolean();
         System.out.println("第"+i+"个随机数: "+flag);
      }
      System.out.println();
      for (int i = 1; i <= 3; i++) {
         double num = random.nextDouble();
         System.out.println("第"+i+"个随机数: "+num);
      }
      System.out.println();
      for (int i = 1; i <= 3; i++) {
         float num = random.nextFloat();
         System.out.println("第"+i+"个随机数: "+num);
      }
      System.out.println();
      for (int i = 1; i <= 3; i++) {
         int num = random.nextInt();
         System.out.println("第"+i+"个随机数: "+num);
      }
      System.out.println();
      for (int i = 1; i <= 3; i++) {
         int num = random.nextInt(10);
```

```
            System.out.println("第"+i+"个随机数: "+num);
        }
        System.out.println();
        for (int i = 1; i <= 3; i++) {
            long num = random.nextLong();
            System.out.println("第"+i+"个随机数: "+num);
        }
        System.out.println();
    }
}
```

运行结果如图 8-4 所示。

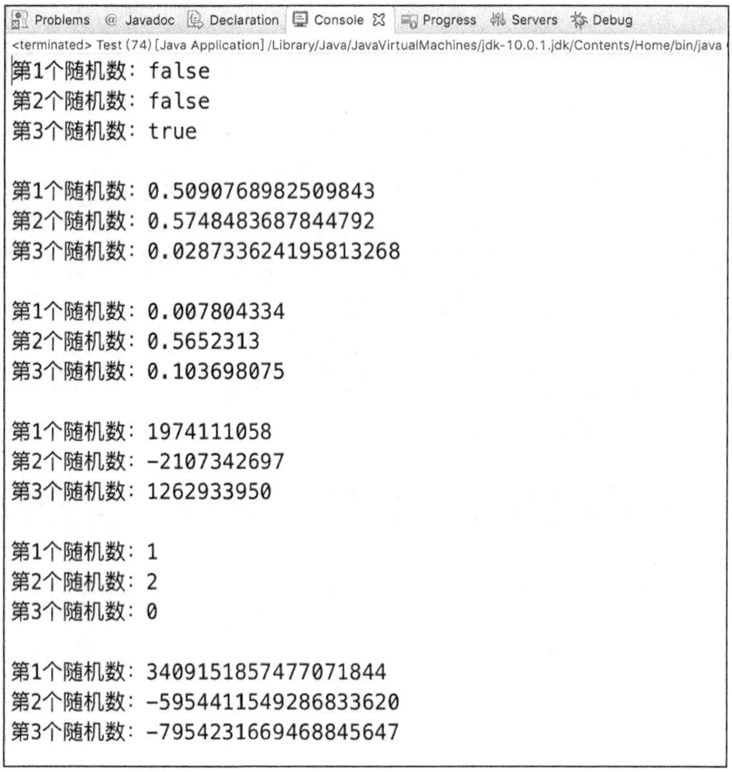
Problems @ Javadoc Declaration Console Progress Servers Debug
<terminated> Test (74) [Java Application] /Library/Java/JavaVirtualMachines/jdk-10.0.1.jdk/Contents/Home/bin/java
第1个随机数: false
第2个随机数: false
第3个随机数: true
第1个随机数: 0.5090768982509843
第2个随机数: 0.5748483687844792
第3个随机数: 0.028733624195813268
第1个随机数: 0.007804334
第2个随机数: 0.5652313
第3个随机数: 0.103698075
第1个随机数: 1974111058
第2个随机数: -2107342697
第3个随机数: 1262933950
第1个随机数: 1
第2个随机数: 2
第3个随机数: 0
第1个随机数: 3409151857477071844
第2个随机数: -5954411549286833620
第3个随机数: -7954231669468845647

图 8-4

8.4 String

String 是我们实际开发中使用频率很高的类，Java 通过 String 类来创建和操作字符串数据。本章就为大家详细讲解 String 类。

8.4.1 String 实例化

String 类对象的实例化方式有两种，第 1 种是直接赋值的方式，第 2 种是通过构造函数创建实例化对象的方式。String 类有一个带参数的构造函数：public String(String original），将 String 对象的值直接传入即可创建。两种方式的具体操作如代码 8-8 所示。

代码 8-8

```
public class StringTest {
   public static void main(String[] args) {
      //第一种方式直接赋值
      String str1 = "Hello";
      //第二种方式通过构造函数
      String str2 = new String("World");
      System.out.println(str1);
      System.out.println(str2);
   }
}
```

运行结果如图 8-5 所示。

图 8-5

两种实例化方式有什么不同呢？哪种方式更优呢？我们先来看下面的代码 8-9。

代码 8-9

```
public class StringTest {
   public static void main(String[] args) {
      String str1 = "Hello";
      String str2 = "Hello";
      System.out.println(str1 == str2);
      String str3 = new String("World");
      String str4 = new String("World");
      System.out.println(str3 == str4);
   }
}
```

上述代码非常简单，用直接赋值的方式创建了 String 对象 str1 和 str2，并且值相等。又用构造函数的方式创建了 String 对象 str3 和 str4，同样值相等。然后用==分别判断 str1 和 str2 是否相等，以及 str3 和 str4 是否相等。==判断的并不是对象的值是否相等，而是判断对象所引用的内存地址是否相等，运行结果如图 8-6 所示。

图 8-6

通过结果我们可以得出结论，str1 所引用的内存地址和 str2 所引用的内存地址相同，即 str1 和 str2 指向堆内存中的同一块位置。而 str3 和 str4 则指向堆内存中不同的位置，那么问题来了，在值相等的情况下，为什么会有截然不同的两种结果呢？

这是因为第一种直接赋值的方式如"String str1 = "Hello""，会首先在栈内存中开辟一块空间来保存变量 str1，同时在堆内存中开辟一块合适的空间来存储"Hello"，然后将堆内存的地址赋给栈内存中的 str1，即 str1 中存储的是"Hello"的内存地址。Java 同时在堆内存中提供了一个字符串常量池，专门用来存储 String 类型的对象。此外字符串常量池有一个特点，在实例化一个 String 对象时，首先会在字符串常量池中查找，如果该字符串已经在池中创建，则直接返回它的内存地址。如果字符串常量池中不存在该字符串，就先创建再返回，即字符串常量池中不会创建值相等的重复字符串对象，str1 和 str2 的创建如图 8-7 所示。

图 8-7

所以 str1 == str2 的返回值是 true，但是字符串常量池只适用于直接赋值的方式。如果是通过构造函数创建的对象则完全不同，这种方式创建的对象会在堆内存中开辟对应的空间来存储。即通过 new String("World") 和 new String("World") 创建了两个对象，虽然值相等，但是会开辟两块内存来存储，所以内存地址肯定是不同的，str3 和 str4 的创建如图 8-8 所示。

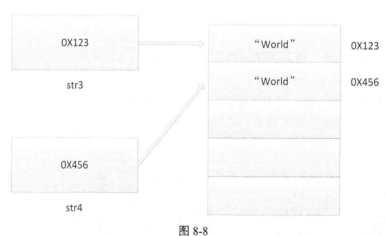

图 8-8

所以 str3 == str4 的返回值为 false，我们在了解了 String 类使用构造函数创建实例化对象的内存模型之后，当判断两个字符串对象是否相同时，就可以直接使用==进行判断了。因为==可以比较内存地址，两个字符串的值无论是否相等，其内存地址肯定不同。那么我们要如何来判断两个字符串对象的值是否相等呢？String 类对继承自 Object 类的 equals()方法进行了重写，如代码 8-10 所示。

代码 8-10

```java
public boolean equals(Object anObject) {
    if (this == anObject) {
        return true;
    }
    if (anObject instanceof String) {
        String aString = (String)anObject;
        if (coder() == aString.coder()) {
            return isLatin1() ? StringLatin1.equals(value, aString.value)
                              : StringUTF16.equals(value, aString.value);
        }
    }
    return false;
}
```

StringLatin1.equals：

```java
public static boolean equals(byte[] value, byte[] other) {
    if (value.length == other.length) {
        for (int i = 0; i < value.length; i++) {
            if (value[i] != other[i]) {
                return false;
            }
        }
        return true;
    }
    return false;
}
```

StringUTF16.equals：

```java
public static boolean equals(byte[] value, byte[] other) {
    if (value.length == other.length) {
        int len = value.length >> 1;
        for (int i = 0; i < len; i++) {
            if (getChar(value, i) != getChar(other, i)) {
                return false;
            }
        }
        return true;
    }
    return false;
}
```

可以看到 String 类在判断两个字符串对象是否相同时，会直接将字符串转为 byte 类型的数组，然后依次判断数组中的每一个值是否相等。如果全部相等，则认为两个字符串相等，返回 true，否则返回 false，表示两个字符串不相同。我们在判断两个字符串对象的值是否相等时，直接调用 String 的 equals()方法即可。

实际上 String 类在存储字符串时，会将字符串的值保存在 byte 类型的数组中，我们知道数组一旦创建，其长度就是不可改变的。既然长度不可改变，也就意味着 byte 类型所存储的字符串值不可修改。一旦修改，就会重新创建一个 String 对象，用新对象的 byte 数组来存储修改之后的字符串。即如果我们修改了 String 对象的值，它就已经不是之前的对象了，而是一个新的对象，如代码 8-11 所示。

代码 8-11

```
public class StringTest {
    public static void main(String[] args) {
        String str1 = new String("Hello");
        String str2 = str1;
        System.out.println(str2 == str1);
        str1 += " World";
        System.out.println(str2 == str1);
    }
}
```

运行结果如图 8-9 所示。

图 8-9

8.4.2　String 常用方法

String 类提供了大量的方法，在实际开发中使用这些方法可以很方便地完成对字符串的操作，常用方法如表 8-2 所示。

表 8-2

方法	描述
public String()	创建一个值为空的对象
public String(String original)	创建一个值为 original 的对象
public String(char value[])	将一个 char 型数组转为字符串对象
public String(char value[], int offset, int count)	将一个指定范围的 char 型数组转为字符串对象
public String(byte[] bytes)	将一个 byte 型数组转为字符串对象
public String(byte bytes[], int offset, int length)	将一个指定范围的 byte 型数组转为字符串对象
public int length()	返回字符串的长度
public boolean isEmpty()	判断字符串是否为空
public char charAt(int index)	返回字符串中指定位置的字符

续表

方法	描述
public byte[] getBytes()	将字符串转为 byte 型数组
public boolean equals(Object anObject)	判断两个字符串是否相等
public boolean equalsIgnoreCase(String anotherString)	判断两个字符串是否相等并且忽略大小写
public int compareTo(String anotherString)	对两个字符串进行排序
public boolean startsWith(String prefix)	判断是否以指定的值开头
public boolean endsWith(String suffix)	判断是否以指定的值结尾
public int hashCode()	获取字符串的散列值
public int indexOf(String str)	从头开始查找指定字符的位置
public int indexOf(String str, int fromIndex)	从指定的位置开始查找指定字符的位置
public String substring(int beginIndex)	截取字符串从指定位置开始到结尾
public String substring(int beginIndex, int endIndex)	截取字符串从指定位置开始到指定位置结束
public String concat(String str)	追加字符串
public String replaceAll(String regex, String replacement)	替换字符串
public String[] split(String regex)	用指定字符串对目标字符串进行分割，返回数组
public String toLowerCase()	将字符串转为小写
public String toUpperCase()	将字符串转为大写
public char[] toCharArray()	将字符串转为 char 型数组

String 常用方法的使用如代码 8-12 所示。

代码 8-12

```java
public class StringTest {
    public static void main(String[] args) {
        char[] array = {'J','a','v','a',' ','H','e','l','l','o',' ','W','o','r','l','d'};
        String str = new String(array);
        System.out.println(str);
        System.out.println("str 长度："+str.length());
        System.out.println("str 是否为空："+str.isEmpty());
        System.out.println("下标为 2 的字符是："+str.charAt(2));
        System.out.println("H 的下标是："+str.indexOf('H'));
        String str2 = "Hello";
        System.out.println("str 和 str2 是否相等："+str.equals(str2));
        String str3 = "HELLO";
        System.out.println("str2 和 str3 忽略大小写是否相等："+str2.equalsIgnoreCase(str3));
        System.out.println("str 是否以 Java 开头："+str.startsWith("Java"));
        System.out.println("str 是否以 Java 结尾："+str.endsWith("Java"));
        System.out.println("从 2 开始截取 str："+str.substring(2));
        System.out.println("从 2 到 6 截取 str："+str.substring(2, 6));
        System.out.println("将 str 中的 World 替换为 Java："+str.replaceAll("World", "Java"));
        System.out.println("用逗号分割 str："+Arrays.toString(str.split(",")));
        System.out.println("将 str 转为 char 类型数组："+Arrays.toString(str.toCharArray()));
        System.out.println("str3 转为小写："+str3.toLowerCase());
        System.out.println("str2 转为大写："+str2.toUpperCase());
    }
}
```

运行结果如图 8-10 所示。

```
<terminated> StringTest [Java Application] /Library/Java/JavaVirtualMachines/jdk-10.0.1.jdk/Contents/Home/bin/java (Oct 14, 2018, 9:08:28 PM)
Java,Hello,World
str长度：16
str是否为空：false
下标为2的字符是：v
H的下标是：5
str和str2是否相等：false
str2和str3忽略大小写是否相等：true
str是否以Java开头：true
str是否以Java结尾：false
从2开始截取str：va,Hello,World
从2到6截取str：va,H
将str中的World替换为Java：Java,Hello,Java
用逗号分割str：[Java, Hello, World]
将str转为char类型数组：[J, a, v, a, ,, H, e, l, l, o, ,, W, o, r, l, d]
str3转为小写：hello
str2转为大写：HELLO
```

图 8-10

8.5 StringBuffer

上一节我们详细介绍了 String 类的使用，在实际开发中使用 String 类会存在一个问题，String 对象一旦创建，其值是不能修改的，如果要修改，会重新开辟内存空间来存储修改之后的对象，即修改了 String 的引用。因为 String 的底层是用数组来存值的，数组长度不可改变这一特性导致了上述问题，所以如果开发中需要对某个字符串进行频繁的修改，使用 String 就不合适了，会造成内存空间的浪费。如何解决这个问题呢？可以使用 StringBuffer 类来解决。StringBuffer 和 String 类似，底层也是用一个数组来存储字符串的值，并且数组的默认长度为 16，即一个空的 StringBuffer 对象，数组长度为 16，如图 8-11 所示。

```java
@HotSpotIntrinsicCandidate
public StringBuffer() {
    super(16);
}
```

图 8-11

当我们调用有参构造创建一个 StringBuffer 对象时，数组长度就不是 16 了，而是根据当前对象的值来决定数组的长度，“值的长度+16”作为数组的长度，如图 8-12 所示。

```java
@HotSpotIntrinsicCandidate
public StringBuffer(String str) {
    super(str.length() + 16);
    append(str);
}
```

图 8-12

所以一个 StringBuffer 创建完成之后，有 16 字节的空间可以对其值进行修改。如果修改的值范围超出了 16 字节，则调用 ensureCapacityInternal()方法继续对底层数组进行扩容，并且保持引用不变，ensureCapacityInternal()方法的具体实现如图 8-13 所示。

```
private void ensureCapacityInternal(int minimumCapacity) {
    // overflow-conscious code
    int oldCapacity = value.length >> coder;
    if (minimumCapacity - oldCapacity > 0) {
        value = Arrays.copyOf(value,
                newCapacity(minimumCapacity) << coder);
    }
}
```

图 8-13

以上就是 StringBuffer 创建及修改的底层原理。接下来，我们来通过代码完成对 StringBuffer 的使用，StringBuffer 常用方法如表 8-3 所示。

表 8-3

方法	描述
public StringBuffer()	无参构造，创建一个空的 StringBuffer
public StringBuffer(String str)	有参构造
public synchronized int length()	返回 StringBuffer 的长度
public synchronized char charAt(int index)	返回字符串中指定位置的字符
public synchronized StringBuffer append(String str)	追加字符
public synchronized StringBuffer delete(int start, int end)	删除指定区间内的字符
public synchronized StringBuffer deleteCharAt(int index)	删除指定位置的字符
public synchronized StringBuffer replace(int start, int end, String str)	将指定区间内的值替换为 str
public synchronized String substring(int start)	截取字符串从指定位置开始到结尾
public synchronized String substring(int start, int end)	截取字符串从指定位置开始到指定位置结束
public synchronized StringBuffer insert(int offset, String str)	向指定位置插入 str
public int indexOf(String str)	从头开始查找指定字符的位置
public int indexOf(String str, int fromIndex)	从指定的位置开始查找指定字符的位置
public synchronized StringBuffer reverse()	进行反转
public synchronized String toString()	返回 StringBuffer 对应的 String

StringBuffer 常用方法的使用如代码 8-13 所示。

代码 8-13

```
public class StringBufferTest {
    public static void main(String[] args) {
        StringBuffer stringBuffer = new StringBuffer();
        System.out.println("StringBuffer:"+stringBuffer);
        System.out.println("StringBuffer 的长度:"+stringBuffer.length());
        stringBuffer = new StringBuffer("Hello World");
        System.out.println("StringBuffer:"+stringBuffer);
        System.out.println("下标为 2 的字符是："+stringBuffer.charAt(2));
        stringBuffer = stringBuffer.append("Java");
        System.out.println("append 之后的 StringBuffer："+stringBuffer);
        stringBuffer = stringBuffer.delete(3, 6);
        System.out.println("delete 之后的 StringBuffer："+stringBuffer);
        stringBuffer = stringBuffer.deleteCharAt(3);
        System.out.println("deleteCharAt 之后的 StringBuffer："+stringBuffer);
        stringBuffer = stringBuffer.replace(2,3,"StringBuffer");
        System.out.println("replace 之后的 StringBuffer："+stringBuffer);
```

```java
            String str = stringBuffer.substring(2);
            System.out.println("substring 之后的 String："+str);
            str = stringBuffer.substring(2,8);
            System.out.println("substring 之后的 String："+str);
            stringBuffer = stringBuffer.insert(6,"six");
            System.out.println("insert 之后的 StringBuffer："+stringBuffer);
            System.out.println("e 的下标是："+stringBuffer.indexOf("e"));
            System.out.println("下标 6 之后的 e 的下标是："+stringBuffer.indexOf("e",6));
            stringBuffer = stringBuffer.reverse();
            System.out.println("reverse 之后的 StringBuffer："+stringBuffer);
            str = stringBuffer.toString();
            System.out.println("StringBuffer 对应的 String："+str);
    }
}
```

运行结果如图 8-14 所示。

```
StringBuffer:
StringBuffer的长度:0
StringBuffer:Hello World
下标为2的字符是: l
append之后的StringBuffer: Hello WorldJava
delete之后的StringBuffer: HelWorldJava
deleteCharAt之后的StringBuffer: HelorldJava
replace之后的StringBuffer: HeStringBufferorldJava
substring之后的String: StringBufferorldJava
substring之后的String: String
insert之后的StringBuffer: HeStrisixngBufferorldJava
e的下标是: 1
下标6之后的e的下标是: 15
reverse之后的StringBuffer: avaJdlroreffuBgnxisirtSeH
StringBuffer对应的String: avaJdlroreffuBgnxisirtSeH
```

图 8-14

8.6 日期类

实际开发中对日期的使用是必不可少的，比如显示系统时间，图书管理系统显示借书日期、还书日期提示等，Java 对日期的使用也提供了良好的封装，主要包括 java.util.Date 和 java.util.Calendar。

8.6.1 Date

Date 类的使用较为简单，直接通过构造函数实例化其对象即可。Date 对象表示当前的系统时间，具体使用如代码 8-14 所示。

代码 8-14

```
public class DateTest {
    public static void main(String[] args) {
        Date date = new Date();
        System.out.println(date);
    }
}
```

运行结果如图 8-15 所示。

图 8-15

通过运行结果可以看到，我们已经获取到了当前的系统时间，但是其表示方式并不符合我们所习惯的日期格式如"2018-10-20"。可以通过 java.text.SimpleDateFormat 类对 Date 对象进行格式化，将日期的表示形式转换成我们所熟悉的方式。我们可以自定义日期的转换格式，SimpleDateFormat 提供了模版标记，如表 8-4 所示。

表 8-4

标记	描述
y	年，yyyy 表示 4 位数的年份信息
M	月，MM 表示 2 位数的月份信息
m	分钟，mm 表示 2 位数的分钟信息
d	天，dd 表示 2 位数的天信息
H	小时，HH 表示 2 位数的 24 小时制下的小时信息
h	小时，hh 表示 2 位数的 12 小时制下的小时信息
s	秒，ss 表示 2 位数的秒信息
S	毫秒，SSS 表示 3 位数的毫秒信息

具体使用如代码 8-15 所示。

代码 8-15

```
public class DateTest {
    public static void main(String[] args) {
        Date date = new Date();
        System.out.println(date);
        //格式化
        SimpleDateFormat simpleDateFormat = new SimpleDateFormat("yyyy-MM-dd HH:mm:ss.SSS");
        String dateStr = simpleDateFormat.format(date);
        System.out.println(dateStr);
    }
}
```

运行结果如图 8-16 所示。

```
Problems  @ Javadoc  Declaration  Console
<terminated> DateTest [Java Application] /Library/Java/JavaVirtualM
Sat Oct 20 16:33:39 CST 2018
2018-10-20 16:33:39.765
```

图 8-16

8.6.2 Calendar

通过 Date 类我们可以获取当前系统时间，但是功能也仅限于此。如果需要对日期数据进行逻辑操作，如计算从当前时间算起 15 天后的日期是几月几号，Date 是没有计算能力的。如果手动来编写逻辑代码又会比较复杂，需要考虑的因素很多，如本月有多少天，如果是 2 月还需要考虑闰月的情况，如果时间很长涉及跨年就更为复杂。每当我们遇到一个比较复杂的功能时，都会发现 Java 已经提供了一个封装好的工具类可以帮我们完成业务代码，这里也不例外，我们可以通过 Calendar 类来完成日期数据的逻辑运算。

使用 Calendar 进行日期运算的基本思路是先将日期数据赋给 Calendar，再调用 Calendar 的方法来完成相关运算，我们首先介绍如何将日期数据赋给 Calendar 类。Calendar 类提供了很多静态常量，用来记录日期数据，常用的静态常量如表 8-5 所示。

表 8-5

常量	描述
public static final int YEAR	年
public static final int MONTH	月
public static final int DAY_OF_MONTH	天，以月为单位，即当天是该月中的第几天
public static final int DAY_OF_YEAR	天，以年为单位，即当天是该年中的第几天
public static final int HOUR_OF_DAY	小时
public static final int MINUTE	分钟
public static final int SECOND	秒
public static final int MILLISECOND	毫秒

Calendar 常用方法的描述如表 8-6 所示。

表 8-6

方法	描述
public static Calendar getInstance()	获取系统对应的 Calendar 实例化对象
public void set(int field, int value)	给静态常量赋值
public int get(int field)	取出静态常量
public final Date getTime()	获取 Calendar 对应的 Date 对象

Calendar 的具体操作如代码 8-16 所示。

代码 8-16

```java
public class CalendarTest {
    public static void main(String[] args) {
        //计算2018年8月6日所在的周是2018年的第几周
        Calendar calendar = Calendar.getInstance();
        calendar.set(Calendar.YEAR, 2018);
        //1月为0，因此8月为7
        calendar.set(Calendar.MONTH, 7);
        calendar.set(Calendar.DAY_OF_MONTH, 6);
        int week = calendar.get(Calendar.WEEK_OF_YEAR);
        System.out.println("2018年8月6日所在的周是2018年的第"+week+"周");

        //计算2018年8月6日往后推21天的日期
        calendar.set(Calendar.DAY_OF_YEAR, calendar.get(Calendar.DAY_OF_YEAR) + 21);
        SimpleDateFormat simpleDateFormat = new SimpleDateFormat("yyyy-MM-dd");
        String laterDateStr = simpleDateFormat.format(calendar.getTime());
        System.out.println("2018年8月6日21天之后的日期："+laterDateStr);

        //计算2018年8月6日往前推21天的日期
        calendar.set(Calendar.YEAR, 2018);
        calendar.set(Calendar.MONTH, 7);
        calendar.set(Calendar.DAY_OF_MONTH, 6);
        calendar.set(Calendar.DAY_OF_YEAR, calendar.get(Calendar.DAY_OF_YEAR) - 21);
        String frontDateStr = simpleDateFormat.format(calendar.getTime());
        System.out.println("2018年8月6日21天之前的日期："+frontDateStr);
    }
}
```

运行结果如图 8-17 所示。

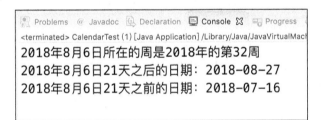

图 8-17

8.7 小结

本章为大家介绍了 Java 中常用的几种实用类，可以把它们理解为 Java 提供给开发者的工具，使用这些工具可以帮助开发者快速完成某些特定功能，而不需要自己编写相关逻辑。这些工具主要包括枚举、Math、Random、String、StringBuffer、Date、Calendar，在使用这些工具类进行开发的同时，建议大家看看这些工具类的源码，学习优秀的编程思想和架构是提高自己编程能力行之有效的方法。

第 9 章 IO 流

在这一章会为大家讲解 Java IO 流的知识，Java IO 操作主要是指使用 Java 程序完成输入（Input）、输出（Output）的功能。所谓输入是指将文件以数据流的形式读取到 Java 程序中，输出是指通过 Java 程序将数据流写入到文件中。输入、输出操作在实际开发中应用较为广泛，例如文件的上传和下载就是通过 IO 流来完成读写的。

9.1 File 类

IO 流可以实现 Java 程序对文件的读写操作，首先我们要掌握的是 Java 如何来操作文件。按照面向对象的编程思想，Java 会用对象来表示文件，文件对象如何创建呢？Java 提供了 java.io.File 类，使用该类的构造函数就可以创建文件对象以表示一个物理资源，File 的常用方法如表 9-1 所示。

表 9-1

方法	描述
public File(String pathname)	根据路径创建对象
public String getName()	获取文件名
public String getParent()	获取文件所在的目录
public File getParentFile()	获取文件所在的目录对应的 File 对象
public String getPath()	获取文件路径
public boolean exists()	判断对象是否存在
public boolean isDirectory()	判断对象是否为目录
public boolean isFile()	判断对象是否为文件
public long length()	获取文件的大小
public boolean createNewFile()	根据当前对象创建新文件
public boolean delete()	删除对象
public boolean mkdir()	根据当前对象创建新目录
public boolean renameTo(File dest)	为已存在的对象重命名

File 常用方法的使用如代码 9-1 所示。

代码 9-1

```
public class FileTest {
```

```java
    public static void main(String[] args) {
        File file = new File("/Users/southwind/Desktop/test.txt");
        boolean flag = file.exists();
        System.out.println("文件是否存在:"+flag);
        String fileName = file.getName();
        System.out.println("文件名:"+fileName);
        long length = file.length();
        System.out.println("文件大小:"+length);
        String path = file.getPath();
        System.out.println("文件路径:"+path);
        String parent = file.getParent();
        System.out.println("文件所在的目录:"+parent);
        File parentFile = file.getParentFile();
        boolean flag2 = parentFile.isDirectory();
        System.out.println("文件的父级对象是否为路径:"+flag2);
        boolean flag3 = parentFile.isFile();
        System.out.println("文件的父级对象是否为文件:"+flag3);
        File file2 = new File("/Users/southwind/Desktop/test2.txt");
        System.out.println("新文件是否存在:"+file2.exists());
        try {
            System.out.println("新文件创建是否成功:"+file2.createNewFile());
        } catch (IOException e) {
            // TODO Auto-generated catch block
            e.printStackTrace();
        }
        File file3 = new File("/Users/southwind/Desktop/test3.txt");
        System.out.println("新文件重命名是否成功:"+file2.renameTo(file3));
        System.out.println("新文件删除是否成功:"+file2.delete());
    }
}
```

运行结果如图 9-1 所示。

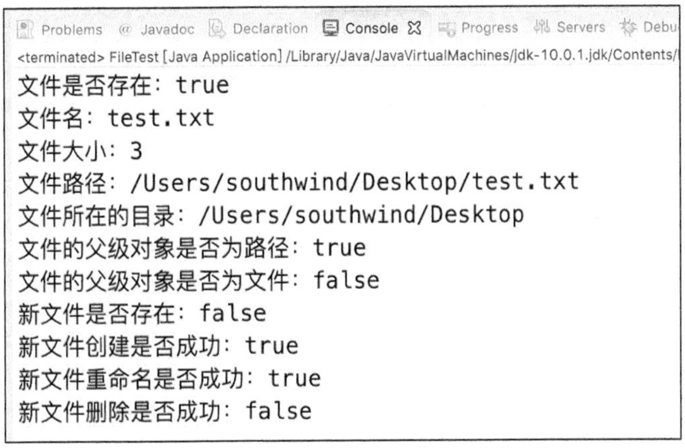

图 9-1

9.2 字节流

了解完 File 对象，接下来我们学习如何读取 File 对象，就是本章的重点内容——IO 流，即通过输入输出流完成对文件的读写。例如，用 Java 程序把 D 盘的文件复制到 F 盘，就可以通过 IO 流来实现。I 是指 input，表示输入的意思；O 是指 Output，表示输出的意思。初学者很容易把这里的输入输出搞混，如何去判断使用输入流还是输出流呢？你只需要站在程序的角度去看待问题，把自己比作程序，首先将 D 盘文件读取到程序中，再由程序将文件的数据流写到 F 盘中。所以读取目标文件就是输入流，写入 F 盘就是输出流，如图 9-2 所示。

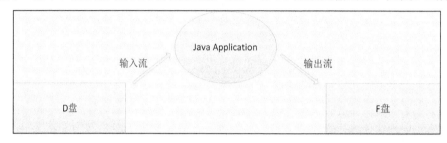

图 9-2

图 9-2 中数据流的走向要搞清楚，流是一种以先进先出的方式传输数据的序列，Java 中的流有很多种不同的分类。

- 按照方向分，可以分为输入流和输出流。
- 按照单位分，可以分为字节流和字符流，字节流指每次处理数据是以字节为单位的，字符流是指每次处理数据是以字符为单位的。
- 按照功能分，可以分为节点流和处理流。

接下来我们分别讲解这几种不同的流，首先来看字节流，它是指以字节为单位来操作数据。打个比方，有一口装满水的水缸和一口空缸，现在要把满缸的水舀到空缸中。舀水需要水瓢，水瓢的大小决定了每次转移水的分量。文件资源就相当于水缸，数据就相当于水，流就相当于水瓢，字节流就是每次只能舀一字节水量的水瓢。

字节流按照方向又可以分为输入字节流（InputStream）和输出字节流（OutputStream），InputStream 是 java.io 包中的顶层父类，并且是一个抽象类，定义如图 9-3 所示。

```
*/
public abstract class InputStream implements Closeable {
```

图 9-3

可以看到 InputStream 实现了 Closeable 接口，该接口的作用是每次操作结束之后完成资源释放，InputStream 常用方法如表 9-2 所示。

表 9-2

方法	描述
public abstract int read() throws IOException	以字节为单位读取数据
public int read(byte b[]) throws IOException	将数据存入 byte 类型数组中，并返回数据长度
public int read(byte b[], int off, int len) throws IOException	将数据存入 byte 类型数组的指定区间中，并返回数据长度
public byte[] readAllBytes() throws IOException	将数据存入 byte 类型数组中返回
public int available() throws IOException	返回当前数据流中未读取的数据个数
public void close() throws IOException	关闭数据流

实际开发中不能直接实例化 InputStream 类的对象，应该实例化其实现了抽象方法的子类 FileInputStream，定义如图 9-4 所示。

```
public
class FileInputStream extends InputStream
{
```

图 9-4

InputStream 的 read() 方法使用如代码 9-2 所示。

代码 9-2

```java
public class IOTest {
    public static void main(String[] args) {
        try {
            InputStream inputStream = new FileInputStream("/Users/southwind/Desktop/test.txt");
            int temp = 0;
            System.out.println("当前未读取的数据个数："+inputStream.available());
            while((temp = inputStream.read())!=-1) {
                System.out.println(temp+",当前未读取的数据个数："+inputStream.available());
            }
            inputStream.close();
            System.out.println("******************************");
            inputStream = new FileInputStream("/Users/southwind/Desktop/test.txt");
            byte[] bytes = new byte[6];
            int length = inputStream.read(bytes);
            System.out.println("数据流长度："+length);
            System.out.println("遍历 byte 数组");
            for (byte b : bytes) {
                System.out.println(b);
            }
            inputStream.close();
            System.out.println("******************************");
            inputStream = new FileInputStream("/Users/southwind/Desktop/test.txt");
            bytes = new byte[10];
            length = inputStream.read(bytes,2,6);
            System.out.println("数据流长度："+length);
            System.out.println("遍历 byte 数组");
            for (byte b : bytes) {
                System.out.println(b);
            }
            inputStream.close();
            System.out.println("******************************");
            inputStream = new FileInputStream("/Users/southwind/Desktop/test.txt");
            bytes = inputStream.readAllBytes();
            System.out.println("遍历 byte 数组");
            for (byte b : bytes) {
                System.out.println(b);
            }
            inputStream.close();
        } catch (FileNotFoundException e) {
        }
    }
}
```

test.txt 的内容如图 9-5 所示。

图 9-5

可以看到 test.txt 中的数据是"ioc"这 3 个字符，转换成 byte 类型的值为 105、111、99，代码 9-2 的运行结果如图 9-6 所示。

```
当前未读取的数据个数：3
105,当前未读取的数据个数：2
111,当前未读取的数据个数：1
99,当前未读取的数据个数：0
****************************
数据流长度：3
遍历byte数组
105
111
99
0
0
0
****************************
数据流长度：3
遍历byte数组
0
0
105
111
99
0
0
0
0
0
****************************
遍历byte数组
105
111
99
```

图 9-6

OutputStream 跟 InputStream 类似，也是一个抽象父类，定义如图 9-7 所示。

```
*/
public abstract class OutputStream implements Closeable, Flushable {
    /**
```

图 9-7

可以看到 OutputStream 除了实现 Closeable 接口之外，还实现了 Flushable 接口。Flushable 接口可以将缓冲区的数据同步到输出流中，保证数据的完整性，OutputStream 的常用方法如表 9-3 所示。

表 9-3

方法	描述
public abstract void write(int b) throws IOException	以字节为单位写数据
public void write(byte b[]) throws IOException	将 byte 类型数组中的数据写出
public void write(byte b[], int off, int len) throws IOException	将 byte 类型数组中指定区间的数据写出
public void flush() throws IOException	可以强制将缓冲区的数据同步到输出流中
public void close() throws IOException	关闭数据流

我们在使用 OutputStream 进行写操作时，不能直接实例化 OutputStream 类的对象，应该实例化其实现了抽象方法的子类 FileOutputStream，定义如图 9-8 所示。

```
public
class FileOutputStream extends OutputStream
{
```

图 9-8

接下来，我们学习对 OutputStream 各种常用方法的使用，write(int b)方法的使用如代码 9-3 所示。

代码 9-3

```
public class IOTest {
    public static void main(String[] args) {
        try {
            OutputStream outputStream = new FileOutputStream("/Users/southwind/Desktop/test2.txt");
            outputStream.write(99);
            outputStream.flush();
            outputStream.close();
        } catch (IOException e) {
        }
    }
}
```

代码 9-3 实现了在桌面创建文件 test2.txt，并且向文件中写入 byte 类型的数据 99。而 byte 类型的数据会转换成字符，99 对应的是'c'，所以运行结果是我们会在桌面看到 test2.txt，如图 9-9 所示。

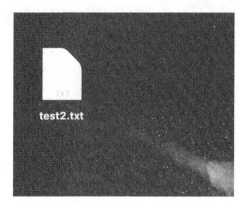

图 9-9

打开 test2.txt，可以看到内容为 c，如图 9-10 所示。

图 9-10

OutputStream 的 write(byte b[])方法的使用如代码 9-4 所示。

代码 9-4

```java
public class IOTest {
    public static void main(String[] args) {
        try {
            OutputStream outputStream = new FileOutputStream("/Users/southwind/Desktop/test2.txt");
            byte[] bytes = {105,111,99};
            outputStream.write(bytes);
            outputStream.flush();
            outputStream.close();
        } catch (IOException e) {
        }
    }
}
```

这次写入的内容是{105,111,99}，转换成字符串为"ioc"，程序运行完成，打开 test2.txt，结果如图 9-11 所示。

图 9-11

OutputStream 的 write(byte b[], int off, int len)方法的使用如代码 9-5 所示。

代码 9-5

```
public class IOTest {
    public static void main(String[] args) {
        try {
            OutputStream outputStream = new FileOutputStream("/Users/southwind/Desktop/test2.txt");
            byte[] bytes = {105,111,99};
            outputStream.write(bytes,1,2);
            outputStream.flush();
            outputStream.close();
        } catch (IOException e) {
        }
    }
}
```

对数组{105，111，99}进行截取，从下标 1 开始取，长度为 2，所以取出的值是"111，99"，转换成字符串为"oc"，程序运行完成后，打开 test2.txt，结果如图 9-12 所示。

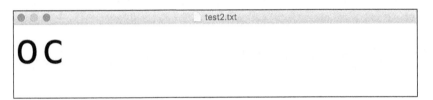

图 9-12

9.3 字符流

字符流是有别于字节流的另外一种数据流，两者的区别在于每次处理的数据单位不同。一个是以字节为单位，一个是以字符为单位，相当于两个尺寸的水瓢，每次舀水的量不同。字符流也分为输入字符流（Reader）和输出字符流（Writer）。本节我们来学习字符流相关的知识。Reader 是一个抽象类，实现了 Readable 和 Closeable 接口，定义如图 9-13 所示。

```
public abstract class Reader implements Readable, Closeable {
```

图 9-13

Readable 接口的作用是可以将数据以字符的形式读入缓冲区，Reader 常用方法如表 9-4 所示。

表 9-4

方法	描述
public int read() throws IOException	以字符为单位读数据
public int read(char cbuf[]) throws IOException	将数据读入 char 类型数组，并返回数据长度

续表

方法	描述
public abstract int read(char cbuf[], int off, int len) throws IOException;	将数据读入 char 类型数组的指定区间,并返回数据长度
public abstract void close() throws IOException	关闭数据流
public long transferTo(Writer out)	将数据直接读入字符输出流

我们在使用 Reader 进行读入操作时,不能直接实例化 Reader 对象,应该实例化其实现了抽象方法的子类 FileReader,定义如图 9-14 所示。

```
public class FileReader extends InputStreamReader {
```

图 9-14

可以看到 FileReader 又继承自 InputStreamReader,这个类才是 Reader 的直接子类,它的作用是将字节流转为字符流,属于处理流。在后面的章节我们会详细讲解 InputStreamReader 类,这里我们直接使用 FileReader 进行实例化操作。接下来我们学习 Reader 的常用方法,还是以操作 test.txt 文件为例,用 Reader 对其进行读取,如代码 9-6 所示。

代码 9-6

```java
public class Test {
    public static void main(String[] args) {
        try {
            Reader reader = new FileReader("/Users/southwind/Desktop/test.txt");
            int temp = 0;
            while((temp = reader.read())!=-1) {
                System.out.println(temp);
            }
            reader.close();
        } catch (FileNotFoundException e) {
        }
    }
}
```

运行结果如图 9-15 所示。

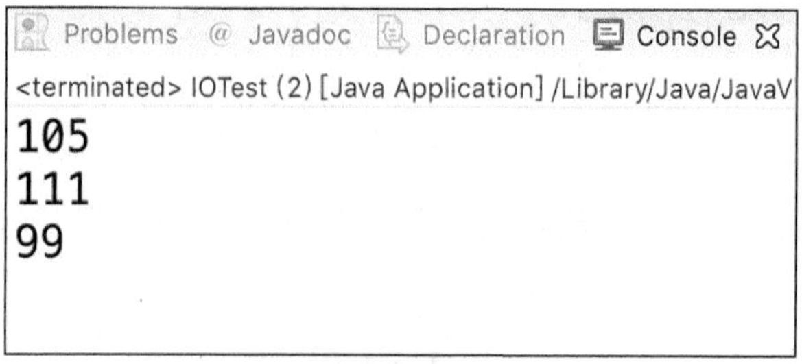

```
<terminated> IOTest (2) [Java Application] /Library/Java/JavaV
105
111
99
```

图 9-15

看到这里,你可能会有疑问了,为什么使用字符流读取出来的结果和使用字节流是一样的?这是因为 test.txt 中保存的数据是 "ioc",在 UTF-8 编码规范下,1 个英文字符占用

1 字节的空间，即这里每读取 1 个字符就等于读取了 1 字节。所以两种流的结果是一样的，Java 工程默认的编码方式为 UTF-8，如图 9-16 所示。

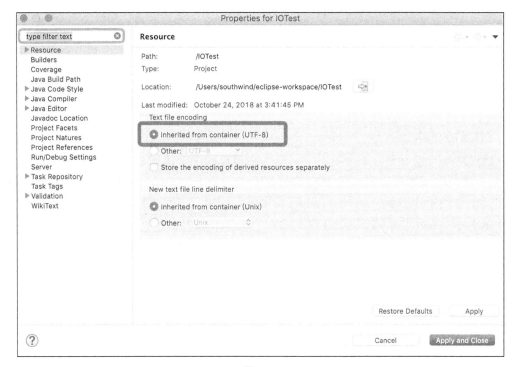

图 9-16

但是汉字就完全不同了，在 UTF-8 编码规范下，1 个汉字占用 3 字节内存。我们也可以将 test.txt 的内容修改为"你好"，如图 9-17 所示。

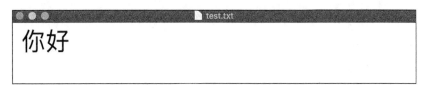

图 9-17

现在使用字符流和字节流同时读取 test.txt，如代码 9-7 所示。

代码 9-7

```java
public class Test {
    public static void main(String[] args) {
        try {
            System.out.println("用字符流读取数据");
            Reader reader = new FileReader("/Users/southwind/Desktop/test.txt");
            int temp = 0;
            while((temp = reader.read())!=-1) {
                System.out.println(temp);
            }
            reader.close();
            System.out.println("用字节流读取数据");
```

```java
            InputStream inputStream = new FileInputStream("/Users/southwind/Desktop/ test.txt");
            int temp2 = 0;
            while((temp2 = inputStream.read())!=-1) {
                System.out.println(temp2);
            }
            inputStream.close();
        } catch (FileNotFoundException e) {
        }
    }
}
```

运行结果如图 9-18 所示。

通过结果可以看到，如果是汉字，那么字符流和字节流读取的结果就完全不同了。"你好"这两个汉字用字符流读取出来是两个字符数据，20320 表示"你"，22909 表示"好"。如果使用字节流，1 个汉字对应 3 字节，所以"228，189，160"这 3 个字节组合起来表示"你"，"229，165，189"这 3 个字节组合起来表示"好"。通过这个例子，相信大家就可以把字节流和字符流的区别搞清楚了，Reader 其他常用方法的使用如代码 9-8 所示。

图 9-18

代码 9-8

```java
public class Test {
    public static void main(String[] args) {
        try {
            Reader reader = new FileReader("/Users/southwind/Desktop/test.txt");
            char[] chars = new char[8];
            int length = reader.read(chars);
            System.out.println("数据流长度："+length);
            System.out.println("遍历 char 数组");
            for (char c : chars) {
                System.out.println(c);
            }
```

```
            reader.close();
            System.out.println("****************************");
            reader = new FileReader("/Users/southwind/Desktop/test.txt");
            chars = new char[8];
            length = reader.read(chars,2,6);
            System.out.println("数据流长度："+length);
            System.out.println("遍历char数组");
            for (char c : chars) {
                System.out.println(c);
            }
            reader.close();
        } catch (FileNotFoundException e) {
        }
    }
}
```

运行结果如图 9-19 所示。

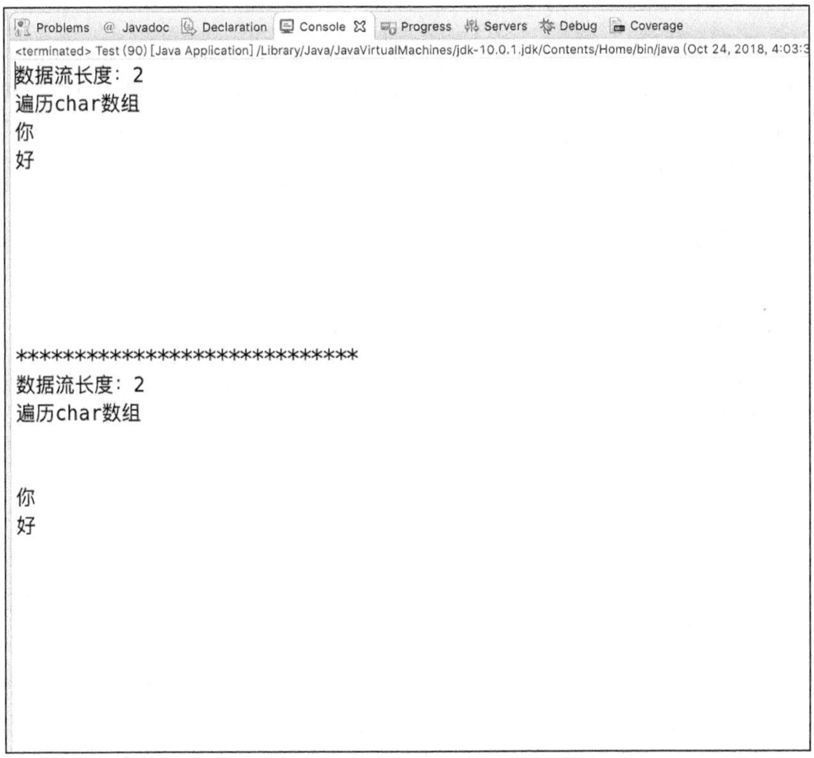

图 9-19

接下来我们学习字符输出流 Writer，定义如图 9-20 所示。

```
public abstract class Writer implements Appendable, Closeable, Flushable {
```

图 9-20

Appendable 接口可以将 char 类型的数据读入数据缓冲区，Writer 的常用方法如表 9-5 所示。

表 9-5

方法	描述
public void write(int c) throws IOException	以字符为单位写数据
public void write(char cbuf[]) throws IOException	将 char 类型数组中的数据写出
public abstract void write(char cbuf[], int off, int len) throws IOException	将 char 类型数组中指定区间的数据写出
public void write(String str) throws IOException	将 String 类型的数据写出
public void write(String str, int off, int len) throws IOException	将 String 类型指定区间的数据写出
public abstract void flush() throws IOException	可以强制将缓冲区的数据同步到输出流中
public abstract void close() throws IOException	关闭数据流

我们在使用 Writer 进行写出操作时，不能直接实例化 Writer 对象，应该实例化其实现了抽象方法的子类 FileWriter，定义如图 9-21 所示。

```
public class FileWriter extends OutputStreamWriter {
```

图 9-21

FileWriter 又继承自 OutputStreamWriter，和 InputStreamReader 一样，OutputStreamWriter 也属于处理流，在后面的章节我们会详细讲解。这里我们直接使用 FileWriter 进行实例化操作，接下来我们学习 Writer 的常用方法如何使用，writer(int c)的使用如代码 9-9 所示。

代码 9-9

```java
public class Test {
    public static void main(String[] args) {
        try {
            Writer writer = new FileWriter("/Users/southwind/Desktop/test2.txt");
            writer.write(20320);
            writer.write(22909);
            writer.flush();
            writer.close();
        } catch (IOException e) {
        }
    }
}
```

代码 9-9 实现了在桌面创建文件 test2.txt，并且向文件中写入 byte 类型的数据 20320 和 22909，byte 类型的数据会转换成对应的字符，20320 对应的是'你'，22909 对应的是'好'，运行结果是我们会看到程序在桌面创建了 test2.txt，如图 9-22 所示。

图 9-22

打开 test2.txt，可以看到内容为你好，如图 9-23 所示。

图 9-23

Writer 的 write(char cbuf[])方法的使用如代码 9-10 所示。

代码 9-10

```java
public class Test {
    public static void main(String[] args) {
        try {
            Writer writer = new FileWriter("/Users/southwind/Desktop/test2.txt");
            char[] chars = {'你','好','世','界'};
            writer.write(chars);
            writer.flush();
            writer.close();
        } catch (IOException e) {
        }
    }
}
```

程序运行完成后，打开 test2.txt，结果如图 9-24 所示。

图 9-24

Writer 的 write(char cbuf[], int off, int len)方法的使用如代码 9-11 所示。

代码 9-11

```java
public class Test {
    public static void main(String[] args) {
        try {
            Writer writer = new FileWriter("/Users/southwind/Desktop/test2.txt");
            char[] chars = {'你','好','世','界'};
            writer.write(chars,2,2);
            writer.flush();
            writer.close();
        } catch (IOException e) {
        }
    }
}
```

程序运行完成后，打开 test2.txt，结果如图 9-25 所示。

图 9-25

Writer 的 write(String str) 方法的使用如代码 9-12 所示。

代码 9-12

```java
public class Test {
    public static void main(String[] args) {
        try {
            Writer writer = new FileWriter("/Users/southwind/Desktop/test2.txt");
            String str = "hello world,你好世界";
            writer.write(str);
            writer.flush();
            writer.close();
        } catch (IOException e) {
        }
    }
}
```

程序运行完成后，打开 test2.txt，结果如图 9-26 所示。

图 9-26

Writer 的 write(String str, int off, int len) 方法的使用如代码 9-13 所示。

代码 9-13

```java
public class Test {
    public static void main(String[] args) {
        try {
            Writer writer = new FileWriter("/Users/southwind/Desktop/test2.txt");
            String str = "hello world,你好世界";
            writer.write(str,10,6);
            writer.flush();
            writer.close();
        } catch (IOException e) {
        }
    }
}
```

程序运行完成后，打开 test2.txt，结果如图 9-27 所示。

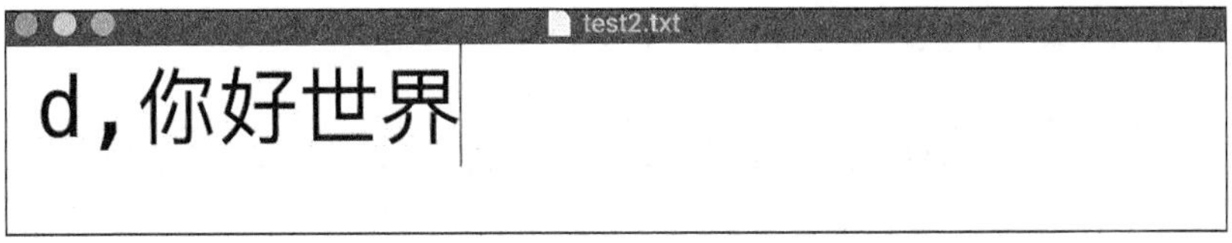

```
                    test2.txt
d,你好世界
```

图 9-27

9.4 处理流

前面的两节我们学习了字节流和字符流，两者的区别在于每次处理数据的单位不同，字节流是基础管道，字符流是在字节流的基础上转换来的。Java 提供了完成转换功能的类，按照输入和输出两个方向分为输入转换流（InputStreamReader）和输出转换流（OutputStreamWriter），我们来看看字符流的类定义，FileReader 的定义如图 9-28 所示。

```
public class FileReader extends InputStreamReader {
```

图 9-28

可以看到，FileReader 继承自 InputStreamReader，即输入转换流。通过这个类将字节流转为字符流，字节是基本单位，所以是基础管道。同理，FileWriter 继承自 OutputStreamWriter，如图 9-29 所示。

```
public class FileWriter extends OutputStreamWriter {
```

图 9-29

InputStreamReader 和 OutputStreamWriter 分别是 Reader 和 Writer 的子类，定义如图 9-30 和图 9-31 所示。

```
public class InputStreamReader extends Reader {
```

图 9-30

```
public class OutputStreamWriter extends Writer {
```

图 9-31

处理流将字节流转换为字符流的过程如图 9-32 所示。

| 文件 | 输入字节流 → | 输入处理流 | 输入字符流 → | 程序 |

图 9-32

InputStreamReader 的使用如代码 9-14 所示。

代码 9-14

```java
public class Test {
    public static void main(String[] args) {
        try {
            InputStream inputStream = new FileInputStream("/Users/southwind/Desktop/test.txt");
            InputStreamReader inputStreamReader = new InputStreamReader(inputStream);
            Reader reader = inputStreamReader;
            char[] chars = new char[1024];
            int length = reader.read(chars);
            reader.close();
            String result = new String(chars,0,length);
            System.out.println(result);
        } catch (FileNotFoundException e) {
        }
    }
}
```

通过上述程序来读取 test.txt 的内容，test.txt 如图 9-33 所示。

图 9-33

程序运行结果如图 9-34 所示。

图 9-34

OutputStreamWriter 的使用如代码 9-15 所示。

代码 9-15

```java
public class Test {
    public static void main(String[] args) {
        String str = "你好 世界";
        try {
            OutputStream outputStream = new FileOutputStream("/Users/southwind/Desktop/test2.txt");
            OutputStreamWriter outputStreamSriter = new OutputStreamWriter(outputStream);
            Writer writer = outputStreamSriter;
```

```
            writer.write(str);
            writer.flush();
            writer.close();
        } catch (FileNotFoundException e) {
        }
    }
}
```

上述程序会将"你好 世界"写入到桌面的 test2.txt 中，运行结果如图 9-35 所示。

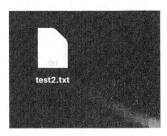

图 9-35

打开 test2.txt，如图 9-36 所示。

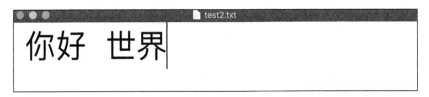

图 9-36

9.5 缓冲流

前面的章节我们已经学习了字节流和字符流，并且对二者的区别进行了介绍。实际上在真实的运行环境中，无论是使用哪种流读取数据都需要频繁地访问硬盘，这样对硬盘的损伤比较大，而且效率不高。为了解决这一问题，减少访问硬盘时的资源开销，我们引入了缓冲流。缓冲流自带缓冲区，可以一次性从硬盘读取部分数据存入缓冲区，再写入程序内存中，如图 9-37 所示。

图 9-37

举个例子，我们把文件和程序内存比作两口水缸，通过字节流或字符流读取数据就相当于用一根管子把一口缸中的水引流到另一口缸。现在我们觉得用管子引流的方式太慢了，那怎么办？我们就用一个大水瓢来舀水，这个大水瓢就是缓冲流，每操作一次可以将很多水存入水瓢中，然后再将整个水瓢中的水倒入另一口缸中。这样就减少了对水缸的访问次数，提高了舀水的效率。

缓冲流也属于处理流，即不能直接关联文件进行操作，只能对已存在的节点流进行包装，如何区分节点流和处理流？如果该流可以直接关联了文件，即提供了一个参数为 File 对象或文件路径的构造函数，该流就是节点流。如果该流不能直接关联文件，即没有提供参数为 File 对象或文件路径的构造函数，而必须依赖其他流来实例化的就是处理流。缓冲流又可以分为字节缓冲流和字符缓冲流，再细分又可分为字节输入缓冲流和字节输出缓冲流，以及字符输入缓冲流和字符输出缓冲流，如图 9-38 所示。

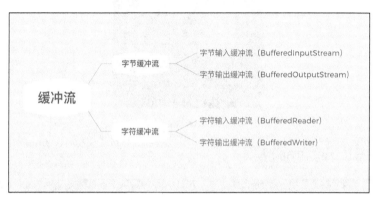

图 9-38

9.5.1 输入缓冲流

首先来看字节输入缓冲流 BufferedInputStream，定义如图 9-39 所示。

```
public
class BufferedInputStream extends FilterInputStream {
```

图 9-39

BufferedInputStream 的常用方法与 OutputStream 类似，并且我们说过缓冲流是处理流，必须依赖于已存在的流才能创建实例化对象，接下来使用 BufferedInputStream 来读取 test.txt，text.txt 的内容如图 9-40 所示。

```
                              test
Java是一门面向对象编程语言，不仅吸收了C++语言的各种优点，
还摒弃了C++里难以理解的多继承、指针等概念，因此Java语言具有功能强大和简单易用两个特征。
Java语言作为静态面向对象编程语言的代表，极好地实现了面向对象理论，
允许程序员以优雅的思维方式进行复杂的编程。
Java具有简单性、面向对象、分布式、健壮性、安全性、平台独立与可移植性、
多线程、动态性等特点。
Java可以编写桌面应用程序、Web应用程序、
分布式系统和嵌入式系统应用程序等。
```

图 9-40

BufferedInputStream 的 read()方法的使用如代码 9-16 所示。

代码 9-16

```java
public class BufferedInputStreamTest {
    public static void main(String[] args) {
        try {
            InputStream inputStream = new FileInputStream("/Users/southwind/Desktop/test.txt");
            BufferedInputStream bufferedInputStream = new BufferedInputStream(inputStream);
            int temp = 0;
            while((temp = bufferedInputStream.read())!=-1) {
                System.out.println(temp);
            }
            bufferedInputStream.close();
            inputStream.close();
        } catch (FileNotFoundException e) {
        }
    }
}
```

运行结果如图 9-41 所示。

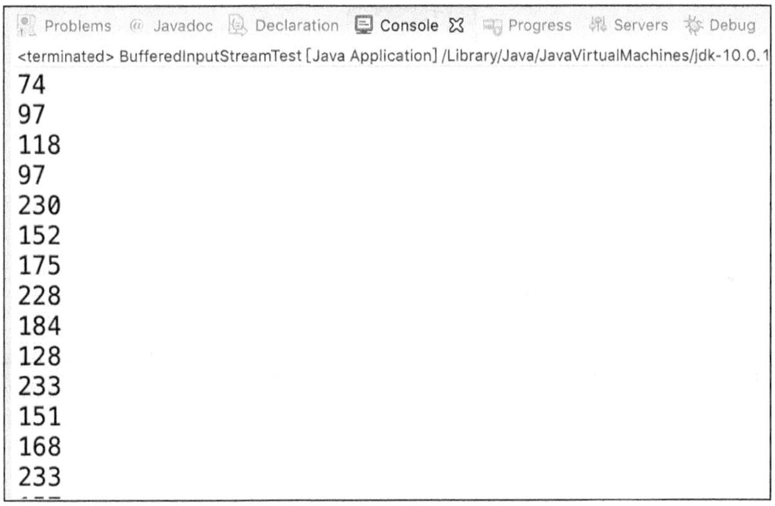

图 9-41

BufferedInputStream 的 read(byte b[])方法的使用如代码 9-17 所示。

代码 9-17

```java
public class BufferedInputStreamTest {
    public static void main(String[] args) {
        try {
            InputStream inputStream = new FileInputStream("/Users/southwind/Desktop/test.txt");
            BufferedInputStream bufferedInputStream = new BufferedInputStream(inputStream);
            int temp = 0;
            byte[] bytes = new byte[1024];
            int length = bufferedInputStream.read(bytes);
            System.out.println(length);
            for (byte b : bytes) {
                System.out.println(b);
```

```
            }
            bufferedInputStream.close();
            inputStream.close();
        } catch (FileNotFoundException e) {
        }
    }
}
```

运行结果如图 9-42 所示。

```
-83
-119
-29
-128
-126
0
0
0
0
0
0
0
0
0
0
```

图 9-42

BufferedInputStream 的 read(byte b[], int off, int len)方法的使用如代码 9-18 所示。

代码 9-18

```
public class BufferedInputStreamTest {
    public static void main(String[] args) {
        try {
            InputStream inputStream = new FileInputStream("/Users/southwind/Desktop/test.txt");
            BufferedInputStream bufferedInputStream = new BufferedInputStream(inputStream);
            int temp = 0;
            byte[] bytes = new byte[1024];
            int length = bufferedInputStream.read(bytes,10,10);
            System.out.println(length);
            for (byte b : bytes) {
                System.out.println(b);
            }
            bufferedInputStream.close();
            inputStream.close();
        } catch (FileNotFoundException e) {
        }
    }
}
```

运行结果如图 9-43 所示。

```
Problems  @ Javadoc  Declaration  Console  Progress  Servers  Debug
<terminated> BufferedInputStreamTest [Java Application] /Library/Java/JavaVirtualMachines/jdk-10.0.1.j
0
0
74
97
118
97
-26
-104
-81
-28
-72
-128
0
0
```

图 9-43

同样是读取 test.txt，接下来我们看字符输入缓冲流 BufferedReader，定义如图 9-44 所示。

```
public class BufferedReader extends Reader {
```

图 9-44

BufferedReader 的 readLine()方法可以直接读取一整行数据，具体使用如代码 9-19 所示。

代码 9-19

```java
public class BufferedReaderTest {
    public static void main(String[] args) {
        try {
            Reader reader = new FileReader("/Users/southwind/Desktop/test.txt");
            BufferedReader bufferedReader = new BufferedReader(reader);
            String str = "";
            while((str = bufferedReader.readLine())!=null) {
                System.out.println(str);
            }
            bufferedReader.close();
            reader.close();
        } catch (FileNotFoundException e) {
        }
    }
}
```

运行结果如图 9-45 所示。

9.5 缓冲流

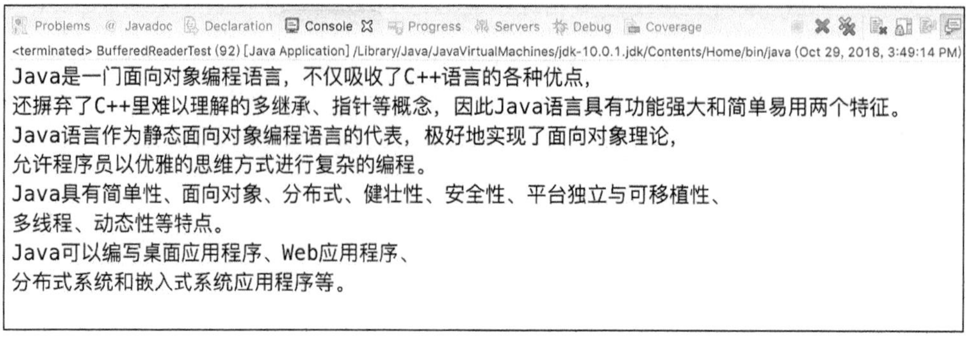

```
图 9-45
```

BufferedReader 的 read()方法的使用如代码 9-20 所示。

代码 9-20

```java
public class BufferedReaderTest {
    public static void main(String[] args) {
        try {
            Reader reader = new FileReader("/Users/southwind/Desktop/test.txt");
            BufferedReader bufferedReader = new BufferedReader(reader);
            int temp = 0;
            while((temp = bufferedReader.read())!=-1) {
                System.out.println(temp);
            }
            bufferedReader.close();
            reader.close();
        } catch (FileNotFoundException e) {
        }
    }
}
```

运行结果如图 9-46 所示。

图 9-46

BufferedReader 的 read(char cbuf[])方法的使用如代码 9-21 所示。

代码 9-21

```java
public class BufferedReaderTest {
    public static void main(String[] args) {
        try {
            Reader reader = new FileReader("/Users/southwind/Desktop/test.txt");
            BufferedReader bufferedReader = new BufferedReader(reader);
            char[] chars = new char[1024];
            int length = bufferedReader.read(chars);
            System.out.println(length);
            for (char c : chars) {
                System.out.println(c);
            }
            bufferedReader.close();
            reader.close();
        } catch (FileNotFoundException e) {
        }
    }
}
```

运行结果如图 9-47 所示。

图 9-47

BufferedReader 的 read(char cbuf[], int off, int len)方法的使用如代码 9-22 所示。

代码 9-22

```java
public class BufferedReaderTest {
    public static void main(String[] args) {
        try {
            Reader reader = new FileReader("/Users/southwind/Desktop/test.txt");
            BufferedReader bufferedReader = new BufferedReader(reader);
            char[] chars = new char[1024];
            int length = bufferedReader.read(chars,10,10);
            System.out.println(length);
```

```
            for (char c : chars) {
                System.out.println(c);
            }
            bufferedReader.close();
            reader.close();
        } catch (FileNotFoundException e) {
        }
    }
}
```

运行结果如图 9-48 所示。

图 9-48

9.5.2 输出缓冲流

了解完输入缓冲流，接下来我们来看输出缓冲流，首先是字节输出缓冲流（BufferedOutputStream），定义如图 9-49 所示。

```
public class BufferedOutputStream extends FilterOutputStream {
```

图 9-49

这次我们通过 Java 程序向文件 test2.txt 中写入数据，BufferedOutputStream 的 write(int b)方法的使用如代码 9-23 所示，需要注意的是，输出缓冲流在关闭之前需要先调用 flush()方法，目的是将缓冲区的数据全部写入到目标文件中，再关闭流。

代码 9-23

```java
public class BufferedOutputStreamTest {
    public static void main(String[] args) {
        try {
            OutputStream outputStream = new FileOutputStream("/Users/southwind/Desktop/test2.txt");
            BufferedOutputStream bufferedOutputStream = new BufferedOutputStream(outputStream);
            String str = "JDK（Java Development Kit）称为Java开发包或Java开发工具，\r\n是一个编写Java的Applet小程序和应用程序的程序开发环境。";
            byte[] bytes = str.getBytes();
            for (byte c : bytes) {
                bufferedOutputStream.write(c);
            }
            bufferedOutputStream.flush();
            bufferedOutputStream.close();
            outputStream.close();
        } catch (FileNotFoundException e) {
        }
    }
}
```

程序运行完成后，test2.txt 的内容如图 9-50 所示。

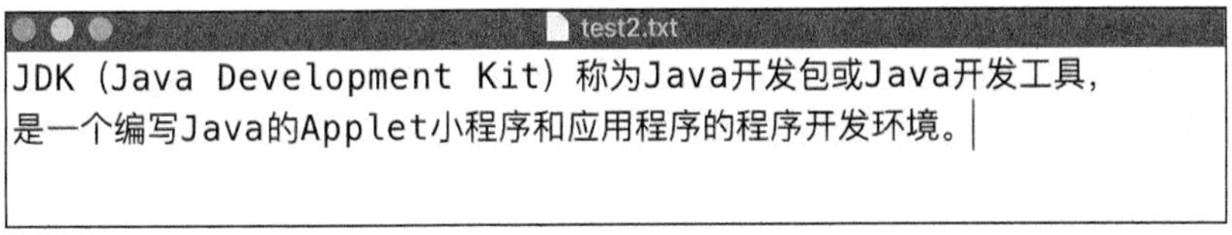

图 9-50

BufferedOutputStream 的 write(byte b[])方法的使用如代码 9-24 所示。

代码 9-24

```java
public class BufferedOutputStreamTest {
    public static void main(String[] args) {
        try {
            OutputStream outputStream = new FileOutputStream("/Users/southwind/Desktop/test2.txt");
            BufferedOutputStream bufferedOutputStream = new BufferedOutputStream(outputStream);
            String str = "JDK（Java Development Kit）称为Java开发包或Java开发工具，\r\n是一个编写Java的Applet小程序和应用程序的程序开发环境。";
            byte[] bytes = str.getBytes();
            bufferedOutputStream.write(bytes);
            bufferedOutputStream.flush();
            bufferedOutputStream.close();
            outputStream.close();
        } catch (FileNotFoundException e) {
        }
    }
}
```

程序运行完成后，test2.txt 的内容如图 9-51 所示。

```
test2.txt
JDK (Java Development Kit) 称为Java开发包或Java开发工具，
是一个编写Java的Applet小程序和应用程序的程序开发环境。
```

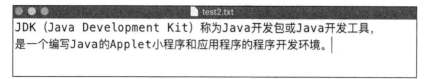

图 9-51

BufferedOutputStream 的 write(byte b[], int off, int len)方法的使用如代码 9-25 所示。

代码 9-25

```
public class BufferedOutputStreamTest {
   public static void main(String[] args) {
      try {
         OutputStream outputStream = new FileOutputStream("/Users/southwind/
Desktop/test2.txt");
         BufferedOutputStream bufferedOutputStream = new BufferedOutputStream
(outputStream);
         String str = "JDK (Java Development Kit) 称为 Java 开发包或 Java 开发工具，\r\n
是一个编写 Java 的 Applet 小程序和应用程序的程序开发环境。";
         byte[] bytes = str.getBytes();
         bufferedOutputStream.write(bytes,10,10);
         bufferedOutputStream.flush();
         bufferedOutputStream.close();
         outputStream.close();
      } catch (FileNotFoundException e) {
      }
   }
}
```

程序运行完成后，test2.txt 的内容如图 9-52 所示。

```
test2.txt
Developme
```

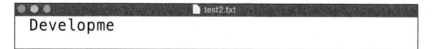

图 9-52

字符输出缓冲流（BufferedWriter），定义如图 9-53 所示。

```
public class BufferedWriter extends Writer {
```

图 9-53

BufferedWriter 提供了一个可以直接将字符串输出的方法 write(String str)，具体使用如代码 9-26 所示。

代码 9-26

```
public class BufferedWriterTest {
```

```java
    public static void main(String[] args) {
        try {
            Writer writer = new FileWriter("/Users/southwind/Desktop/test2.txt");
            BufferedWriter bufferedWriter = new BufferedWriter(writer);
            String str = "JDK (Java Development Kit) 称为 Java 开发包或 Java 开发工具,\r\n是一个编写 Java 的 Applet 小程序和应用程序的程序开发环境。";
            bufferedWriter.write(str);
            bufferedWriter.flush();
            bufferedWriter.close();
            writer.close();
        } catch (IOException e) {
        }
    }
}
```

程序运行完成后，test2.txt 的内容如图 9-54 所示。

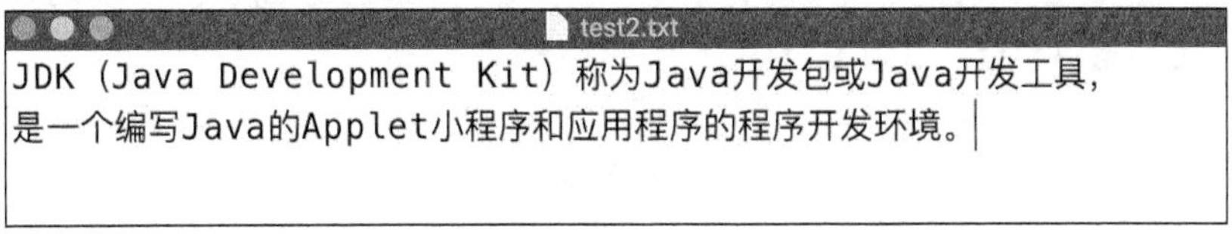

图 9-54

BufferedWriter 的 write(String s, int off, int len)方法的使用如代码 9-27 所示。

代码 9-27

```java
public class BufferedWriterTest {
    public static void main(String[] args) {
        try {
            Writer writer = new FileWriter("/Users/southwind/Desktop/test2.txt");
            BufferedWriter bufferedWriter = new BufferedWriter(writer);
            String str = "JDK (Java Development Kit) 称为 Java 开发包或 Java 开发工具,\r\n是一个编写 Java 的 Applet 小程序和应用程序的程序开发环境。";
            bufferedWriter.write(str, 3, 10);
            bufferedWriter.flush();
            bufferedWriter.close();
            writer.close();
        } catch (IOException e) {
        }
    }
}
```

程序运行完成后，test2.txt 的内容如图 9-55 所示。

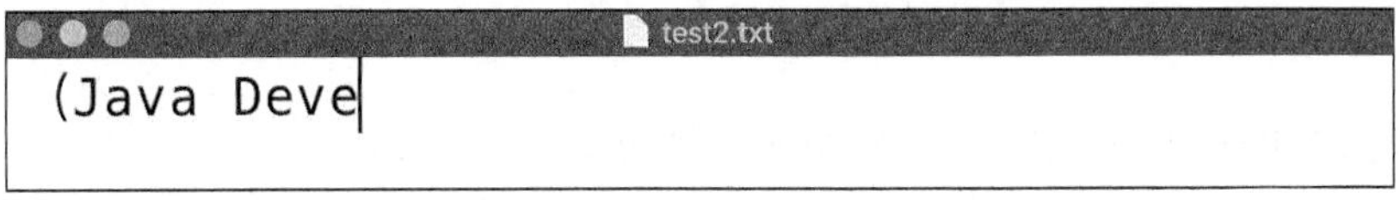

图 9-55

BufferedWriter 的 write(char cbuf[])方法的使用如代码 9-28 所示。

代码 9-28

```java
public class BufferedWriterTest {
    public static void main(String[] args) {
        try {
            Writer writer = new FileWriter("/Users/southwind/Desktop/test2.txt");
            BufferedWriter bufferedWriter = new BufferedWriter(writer);
            char[] chars = {'I',' ','L','o','v','e',' ','J','a','v','a'};
            bufferedWriter.write(chars);
            bufferedWriter.flush();
            bufferedWriter.close();
            writer.close();
        } catch (IOException e) {
        }
    }
}
```

程序运行完成后，test2.txt 的内容如图 9-56 所示。

图 9-56

BufferedWriter 的 write(char cbuf[], int off, int len)方法的使用如代码 9-29 所示。

代码 9-29

```java
public class BufferedWriterTest {
    public static void main(String[] args) {
        try {
            Writer writer = new FileWriter("/Users/southwind/Desktop/test2.txt");
            BufferedWriter bufferedWriter = new BufferedWriter(writer);
            char[] chars = {'I',' ','L','o','v','e',' ','J','a','v','a'};
            bufferedWriter.write(chars,3,6);
            bufferedWriter.flush();
            bufferedWriter.close();
            writer.close();
        } catch (IOException e) {
        }
    }
}
```

程序运行完成后，test2.txt 的内容如图 9-57 所示。

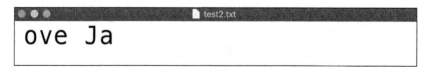

图 9-57

BufferedWriter 的 write(int c)方法的使用如代码 9-30 所示。

代码 9-30

```java
public class BufferedWriterTest {
    public static void main(String[] args) {
```

```
        try {
            Writer writer = new FileWriter("/Users/southwind/Desktop/test2.txt");
            BufferedWriter bufferedWriter = new BufferedWriter(writer);
            bufferedWriter.write(22909);
            bufferedWriter.flush();
            bufferedWriter.close();
            writer.close();
        } catch (IOException e) {
        }
    }
}
```

程序运行完成后，test2.txt 的内容如图 9-58 所示。

图 9-58

9.6 序列化和反序列化

前面我们介绍的所有流都是将文件读取到内存中，以 byte、数组、char 或者 String 类型展示的，同理我们也可以将内存中的数字或者字符串数据输出到文件中。那么如果我们需要将内存中的对象输出到文件中，或者从文件中读取数据并还原成内存中的对象，应该如何处理呢？

这就是本节要学习的内容：序列化和反序列化。序列化就是指将内存中的对象输出到硬盘文件中进行保存，反序列化就是相反的操作，从文件中读取数据并还原成内存中的对象。

9.6.1 序列化

我们先来看序列化，首先给需要序列化的对象的类实现 java.io.Serializable 接口，如代码 9-31 所示。

代码 9-31

```
public class User implements Serializable{
    private int id;
    private String name;
    private int age;
    private char gender;
     //getter、setter方法
}
```

接下来就可以将 User 的实例化对象进行序列化处理，通过数据流写入到文件中，这里需要用到 ObjectOutputStream，该类是专门将对象进行输出处理的，定义如图 9-59 所示。

```
public class ObjectOutputStream
    extends OutputStream implements ObjectOutput, ObjectStreamConstants
```

图 9-59

ObjectOutputStream 的使用如代码 9-32 所示。

代码 9-32

```java
public class SerializableTest {
    public static void main(String[] args) {
        User user = new User();
        user.setId(1);
        user.setName("张三");
        user.setGender('男');
        user.setAge(22);
        try {
            OutputStream outputStream = new FileOutputStream("/Users/southwind/Desktop/test2.txt");
            ObjectOutputStream objectOutputStream = new ObjectOutputStream(outputStream);
            objectOutputStream.writeObject(user);
            objectOutputStream.flush();
            objectOutputStream.close();
            outputStream.close();
        } catch (FileNotFoundException e) {
        }
    }
}
```

程序运行完成后，test2.txt 的内容如图 9-60 所示。

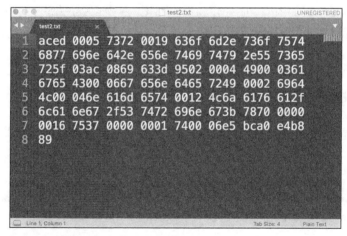

图 9-60

9.6.2 反序列化

图 9-60 显示的数据就是 User 对象被序列化保存到文件中的值。接下来完成反序列的操作，即从文件中读取数据并还原成内存中的对象，这里需要用到 ObjectInputStream 类，

定义如图 9-61 所示。

```
public class ObjectInputStream
    extends InputStream implements ObjectInput, ObjectStreamConstants
```

图 9-61

ObjectInputStream 的使用如代码 9-33 所示。

代码 9-33

```java
public class SerializableTest {
    public static void main(String[] args) {
        try {
            InputStream inputStream = new FileInputStream("/Users/southwind/Desktop/test2.txt");
            ObjectInputStream objectInputStream = new ObjectInputStream(inputStream);
            User user = (User)objectInputStream.readObject();
            objectInputStream.close();
            inputStream.close();
            System.out.println(user.getId());
            System.out.println(user.getName());
            System.out.println(user.getGender());
            System.out.println(user.getAge());
        } catch (FileNotFoundException e) {
        }
    }
}
```

运行结果如图 9-62 所示。

```
<terminated> SerializableTest (1) [Java Application] /Library/Java/JavaV
1
张三
男
22
```

图 9-62

9.7 小结

本章为大家讲解了 Java 中的 IO 流，这是很重要的一章，因为在实际的项目开发中 IO 流的使用较为广泛，比如文件上传、下载的底层就是通过 IO 流来实现的，再比如要读取一些配置文件也是通过 IO 流来完成的。Java 中 IO 流的体系结构较为庞大，根据单位、方向和功能可分为多种不同的数据流，并且这些分类中也存在交叉。作为初学者首先应该搞清楚流的分类，例如什么是输出流，什么是输入流，什么时候使用字节流，什么时候使用字符流等，搞清楚这些问题之后，应用起来才会更加得心应手。

第4部分 底层扩展

第 10 章 反射

> 前面的 9 章内容对 Java 的基本语法、面向对象编程思想和常用类库进行了详细的讲解。本章我们来学习 Java 的反射机制，反射是重点也是难点，是 Java 中非常重要的知识点，应用面很广，大部分类库以及框架底层都用到了反射机制，它是动态语言的关键。反射是一个比较抽象的概念，不好理解，这里先简单做一个解释，让大家对它有一个基本的认识和概念，后续的章节再来详细阐述。反射顾名思义就是反转执行，我们日常生活中的反射有哪些呢？比如通过照镜子可以反射出你的容貌，水面可以反射出物体的形态，等等。无论是哪种反射，都可以通过一个虚像映射到实物，这样我们就可以获取实物的某些形态特征。
>
> 程序中的反射也是同样的道理，它完成的是通过一个实例化对象映射到类，这样我们在程序运行期间就可以获取类的信息了。看到这里大家可能会有疑问，类的结构都是开发者自定义的，它的信息我完全可以看到啊，为什么还要通过反射来获取呢？注意这里讲的通过反射获取类信息是在程序运行期间，我们可以直接看到的类结构是静态的，而程序运行起来是动态的，两者是不同的两个概念。我们要做的就是在运行期间获取类的结构然后完成某些特定功能。一句话来简单理解反射：常规情况下是通过类来创建实例化对象的，反射就是将这一过程进行反转，通过实例化对象来获取对应的类信息。

10.1 Class 类

　　Class 类是反射的基础，上文提过反射就是通过对象来获取类的信息。那么问题来了，Java 程序中如何来描述类的信息？或者类的信息应该以一种什么样的形式呈现呢？我们知道 Java 是面向对象的编程语言，Java 世界中的一切都可以看作对象，这里也不例外，可以将通过反射获取的类信息抽象成对象。

　　既然是对象，那么就一定有对应的类，这个类就是 Class。所以 Class 这个类你可以把它理解为专门用来描述其他类的类，Class 的每一个实例对象对应的都是其他类的结构特征，定义如图 10-1 所示。

```
public final class Class<T> implements java.io.Serializable,
                                       GenericDeclaration,
                                       Type,
                                       AnnotatedElement {
```

图 10-1

在外部不能通过构造函数来实例化 Class 对象，因为 Class 类只有一个私有的构造函数，如图 10-2 所示。

```
private Class(ClassLoader loader, Class<?> arrayComponentType) {
    // Initialize final field for classLoader.  The initialization value of non-null
    // prevents future JIT optimizations from assuming this final field is null.
    classLoader = loader;
    componentType = arrayComponentType;
}
```

图 10-2

我们要如何获取 Class 的实例化对象来描述其他类的结构呢？既然 Class 的实例化对象是描述其他类的，那么在创建 Class 对象时就需要用到被描述的类，创建 Class 实例化对象的方式有 3 种。

方式 1：调用 Class 的静态方法 forName(String className) 创建，将目标类的全限定类名（全限定类名就是包含所在包信息的类名全称，如 java.lang.String）作为参数传入，即可获取对应的 Class 对象，forName(String className) 方法的定义如图 10-3 所示。

```
@CallerSensitive
public static Class<?> forName(String className)
            throws ClassNotFoundException {
    Class<?> caller = Reflection.getCallerClass();
    return forName0(className, true, ClassLoader.getClassLoader(caller), caller);
}
```

图 10-3

方式 2：通过目标类的 class 创建，Java 中的每一个类都可以调用类.class，这里的 class 不是属性，它叫作"类字面量"，其作用是获取内存中目标类型 class 对象的引用。

方式 3：通过目标类实例化对象的 getClass() 方法创建。getClass() 方法定义在 Object 类中，被所有类继承，通过实例对象获取内存中该类 class 对象的引用，getClass() 方法的定义如图 10-4 所示。

```
@HotSpotIntrinsicCandidate
public final native Class<?> getClass();
```

图 10-4

上述 3 种创建 Class 实例对象的具体操作如代码 10-1 所示。

代码 10-1

```java
public class Test {
    public static void main(String[] args) {
        try {
            Class clazz = Class.forName("java.lang.String");
```

```
            System.out.println(clazz);
        } catch (ClassNotFoundException e) {
        }
        Class clazz2 = String.class;
        System.out.println(clazz2);
        String str = new String("Hello");
        Class clazz3 = str.getClass();
        System.out.println(clazz3);
    }
}
```

运行结果如图 10-5 所示。

```
class java.lang.String
class java.lang.String
class java.lang.String
```

图 10-5

我们可以看到在结果中打印了 3 次 "class java.lang.String"，为什么会是这样的结果呢？因为在打印某个对象时，会自动调用它的 toString()方法将该对象转为一个字符串。toString()是定义在 Object 类中的方法，所有的类都可以继承并进行重写，很显然 Class 类就对 toString()方法进行了重写，如图 10-6 所示。

```
public String toString() {
    return (isInterface() ? "interface " : (isPrimitive() ? "" : "class "))
        + getName();
}
```

图 10-6

在上述代码中我们获取了 3 个 Class 对象，那么这 3 个 Class 对象是否相等，即在内存中到底是创建了一个 Class 对象还是 3 个 Class 对象呢？我们通过代码来验证，修改上述代码如代码 10-2 所示。

代码 10-2

```
public class Test {
    public static void main(String[] args) {
        Class clazz = null;
        try {
            clazz = Class.forName("java.lang.String");
            System.out.println(clazz);
        } catch (ClassNotFoundException e) {
        }
        Class clazz2 = String.class;
        System.out.println(clazz2);
        String str = new String("Hello");
        Class clazz3 = str.getClass();
```

```
            System.out.println(clazz3);
            System.out.println(clazz.hashCode());
            System.out.println(clazz2.hashCode());
            System.out.println(clazz3.hashCode());
            System.out.println(clazz == clazz2);
            System.out.println(clazz2 == clazz3);
        }
    }
```

运行结果如图 10-7 所示。

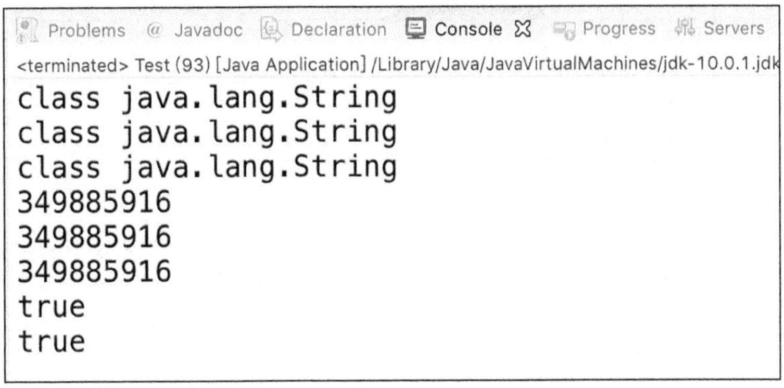

```
class java.lang.String
class java.lang.String
class java.lang.String
349885916
349885916
349885916
true
true
```

图 10-7

通过结果可以看到 3 个 Class 对象的散列值相同，同时依次比较 clazz 和 clazz2，clazz2 和 clazz3 的内存地址也相同，证明这 3 个 Class 对象是相等的，但内存中只有一份。这一点很好理解，因为 Class 的实例对象是用来描述某个目标类的，而每一个目标类的运行时类在内存中只有一份，所以对应的 Class 对象也只有一份。

10.2 获取类结构

上一节我们讲到了如何获取目标类对应的 Class 对象，Class 是整个反射机制的源头，跟反射相关的操作大部分是基于对 Class 对象的操作，即获取目标类的信息都是通过调用 Class 的相关方法来完成的。类的信息包括其内部的成员变量、方法、构造函数、继承的父类和实现的接口等，本节我们就来逐一讲解如何获取类的这些信息，Class 类的常用方法如表 10-1 所示。

表 10-1

方法	描述
public native boolean isInterface()	判断该类是否为接口
public native boolean isArray()	判断该类是否为数组
public boolean isAnnotation()	判断该类是否为注解
public String getName()	获取该类的全限定类名
public ClassLoader getClassLoader()	获取类加载器
public native Class<? super T> getSuperclass()	获取该类的直接父类

续表

方法	描述
public Package getPackage()	获取该类所在的包
public String getPackageName()	获取该类所在包的名称
public Class<?>[] getInterfaces()	获取该类的全部接口
public native int getModifiers()	获取该类的访问权限修饰符
public Field[] getFields() throws SecurityException	获取该类的全部公有成员变量，包括继承自父类的和自定义的
public Field[] getDeclaredFields() throws SecurityException	获取该类的自定义成员变量
public Field getField(String name) throws NoSuchFieldException, SecurityException	通过名称获取该类的公有成员变量，包括继承自父类的和自定义的
public Field getDeclaredField(String name) throws NoSuchFieldException, SecurityException	通过名称获该类的自定义成员变量
public Method[] getMethods() throws SecurityException	获取该类的全部公有方法，包括继承自父类的和自定义的
public Method[] getDeclaredMethods() throws SecurityException	获取该类的自定义方法
public Method getMethod(String name, Class<?>... parameterTypes) throws NoSuchMethodException, SecurityException	通过名称和参数信息获取该类的公有方法，包括继承自父类的和自定义的
public Method getDeclaredMethod(String name, Class<?>... parameterTypes) throws NoSuchMethodException, SecurityException	通过名称和参数信息获取该类的自定义方法
public Constructor<?>[] getConstructors() throws SecurityException	获取该类的公有构造函数
public Constructor<?>[] getDeclaredConstructors() throws SecurityException	获取该类的全部构造函数
public Constructor<T> getConstructor(Class<?>... parameterTypes) throws NoSuchMethodException, SecurityException	通过参数信息获取该类的公有构造函数
public Constructor<T> getDeclaredConstructor(Class<?>... parameterTypes) throws NoSuchMethodException, SecurityException	通过参数信息获取该类的构造函数

接下来我们一起学习这些常用方法。

10.2.1 获取类的接口

Class 提供了 getInterfaces()方法来获取目标类的接口，该方法定义如图 10-8 所示。

```
public Class<?>[] getInterfaces() {
    // defensively copy before handing over to user code
    return getInterfaces(true);
}

private Class<?>[] getInterfaces(boolean cloneArray) {
    ReflectionData<T> rd = reflectionData();
    if (rd == null) {
        // no cloning required
        return getInterfaces0();
    } else {
        Class<?>[] interfaces = rd.interfaces;
        if (interfaces == null) {
            interfaces = getInterfaces0();
            rd.interfaces = interfaces;
        }
        // defensively copy if requested
        return cloneArray ? interfaces.clone() : interfaces;
    }
}
```

图 10-8

可以看到，getInterfaces()方法的返回值是一个 Class 类型的数组，首先为什么是数组呢？因为一个类是可以同时实现多个接口的，所以需要用数组来保存返回值，表示一个或多个。数组的类型为什么是 Class 呢？因为这里同样是用对象来描述接口的信息，所以返回的是 Class 类型的对象。getInterfaces()方法的具体使用如代码 10-3 所示，首先自定义 Student 类并实现 Serializable 和 Comparable<T>接口，然后通过反射来获取 Student 类的接口信息。

代码 10-3

```java
public class Student implements Serializable,Comparable<Student>{
   @Override
   public int compareTo(Student o) {
      // TODO Auto-generated method stub
      return 0;
   }
}


public class Test {
   public static void main(String[] args) {
      try {
         Class clazz = Class.forName("com.southwind.entity.Student");
         Class[] interfaces = clazz.getInterfaces();
         for (Class class1 : interfaces) {
            System.out.println(class1);
         }
      } catch (ClassNotFoundException e) {
      }
   }
}
```

运行结果如图 10-9 所示。

```
Problems  @ Javadoc  Declaration  Console  Progress
<terminated> Test (93) [Java Application] /Library/Java/JavaVirtualMachines/
interface java.io.Serializable
interface java.lang.Comparable
```

图 10-9

10.2.2 获取父类

Class 提供了 getSuperclass()方法来获取目标类的父类，该方法定义如图 10-10 所示。

```
@HotSpotIntrinsicCandidate
public native Class<? super T> getSuperclass();
```

图 10-10

getSuperclass 是一个本地方法，为 Student 创建父类 People，然后通过反射机制来获取 Student 类的父类信息，具体实现如代码 10-4 所示。

代码 10-4

```java
public class Student extends People{
}


public class Test {
   public static void main(String[] args) {
      try {
         Class clazz = Class.forName("com.southwind.entity.Student");
         Class superClass = clazz.getSuperclass();
         System.out.println("Student 的父类是："+superClass);
         clazz = Class.forName("com.southwind.entity.People");
         superClass = clazz.getSuperclass();
         System.out.println("People 的父类是："+superClass);
      } catch (ClassNotFoundException e) {
      }
   }
}
```

运行结果如图 10-11 所示。

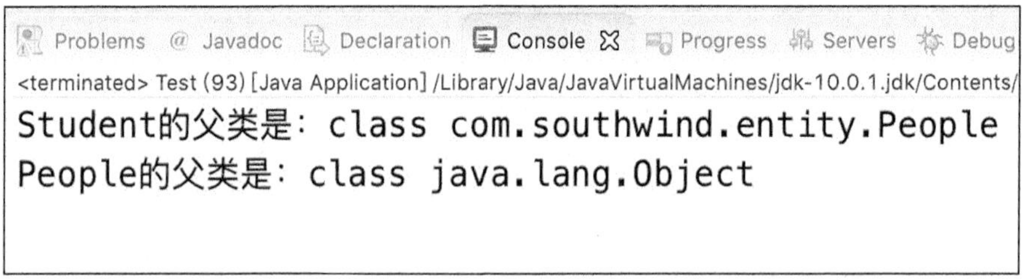

```
Student的父类是: class com.southwind.entity.People
People的父类是: class java.lang.Object
```

图 10-11

10.2.3 获取构造函数

Class 提供了 4 个方法获取构造函数，分别是 getConstructor(Class<?>... parameterTypes)、getConstructors()、getDeclaredConstructor(Class<?>...parameterTypes) 和 getDeclaredConstructors()，各个方法的定义如图 10-12～图 10-15 所示。

```java
@CallerSensitive
public Constructor<T> getConstructor(Class<?>... parameterTypes)
    throws NoSuchMethodException, SecurityException
{
    SecurityManager sm = System.getSecurityManager();
    if (sm != null) {
        checkMemberAccess(sm, Member.PUBLIC, Reflection.getCallerClass(), true);
    }
    return getReflectionFactory().copyConstructor(
        getConstructor0(parameterTypes, Member.PUBLIC));
}
```

图 10-12

```
@CallerSensitive
public Constructor<?>[] getConstructors() throws SecurityException {
    SecurityManager sm = System.getSecurityManager();
    if (sm != null) {
        checkMemberAccess(sm, Member.PUBLIC, Reflection.getCallerClass(), true);
    }
    return copyConstructors(privateGetDeclaredConstructors(true));
}
```

图 10-13

```
@CallerSensitive
public Constructor<T> getDeclaredConstructor(Class<?>... parameterTypes)
    throws NoSuchMethodException, SecurityException
{
    SecurityManager sm = System.getSecurityManager();
    if (sm != null) {
        checkMemberAccess(sm, Member.DECLARED, Reflection.getCallerClass(), true);
    }

    return getReflectionFactory().copyConstructor(
        getConstructor0(parameterTypes, Member.DECLARED));
}
```

图 10-14

```
@CallerSensitive
public Constructor<?>[] getDeclaredConstructors() throws SecurityException {
    SecurityManager sm = System.getSecurityManager();
    if (sm != null) {
        checkMemberAccess(sm, Member.DECLARED, Reflection.getCallerClass(), true);
    }
    return copyConstructors(privateGetDeclaredConstructors(false));
}
```

图 10-15

具体使用如代码 10-5 所示。

代码 10-5

```java
public class Student extends People{
    public Student() {
    }
    private Student(int id) {
    }
}


public class Test {
    public static void main(String[] args) {
        try {
            Class clazz = Class.forName("com.southwind.entity.Student");
            Constructor[] constructors = clazz.getConstructors();
            System.out.println("***************公有构造函数***************");
            for (Constructor constructor : constructors) {
                System.out.println(constructor);
            }
            System.out.println("***************全部构造函数***************");
            Constructor[] declaredConstructors = clazz.getDeclaredConstructors();
            for (Constructor constructor : declaredConstructors) {
                System.out.println(constructor);
```

```
            }
            System.out.println("***************公有无参构造函数***************");
            Constructor constructor = clazz.getConstructor(null);
            System.out.println(constructor);
            System.out.println("***************私有带参构造函数***************");
            Constructor declaredConstructor = clazz.getDeclaredConstructor(int.class);
            System.out.println(declaredConstructor);
        } catch (ClassNotFoundException e) {
        }
    }
}
```

运行结果如图 10-16 所示。

```
Problems @ Javadoc Declaration Console ☒ Progress Servers Debug
<terminated> Test (93) [Java Application] /Library/Java/JavaVirtualMachines/jdk-10.0.1.jdk/Contents/H
***************公有构造函数***************
public com.southwind.entity.Student()
***************全部构造函数***************
public com.southwind.entity.Student()
private com.southwind.entity.Student(int)
***************公有无参构造函数***************
public com.southwind.entity.Student()
***************私有带参构造函数***************
private com.southwind.entity.Student(int)
```

图 10-16

10.2.4 获取方法

Class 提供了 4 个方法获取类中的方法，分别是 getMethod(String name, Class<?>... parameterTypes)、getMethods()、getDeclaredMethod(String name, Class<?>... parameterTypes) 和 getDeclaredMethods()，各个方法的定义如图 10-17～图 10-20 所示。

```
@CallerSensitive
public Method getMethod(String name, Class<?>... parameterTypes)
    throws NoSuchMethodException, SecurityException {
    Objects.requireNonNull(name);
    SecurityManager sm = System.getSecurityManager();
    if (sm != null) {
        checkMemberAccess(sm, Member.PUBLIC, Reflection.getCallerClass(), true);
    }
    Method method = getMethod0(name, parameterTypes);
    if (method == null) {
        throw new NoSuchMethodException(methodToString(name, parameterTypes));
    }
    return getReflectionFactory().copyMethod(method);
}
```

图 10-17

```
@CallerSensitive
public Method[] getMethods() throws SecurityException {
    SecurityManager sm = System.getSecurityManager();
    if (sm != null) {
        checkMemberAccess(sm, Member.PUBLIC, Reflection.getCallerClass(), true);
    }
    return copyMethods(privateGetPublicMethods());
}
```

图 10-18

```
@CallerSensitive
public Method getDeclaredMethod(String name, Class<?>... parameterTypes)
    throws NoSuchMethodException, SecurityException {
    Objects.requireNonNull(name);
    SecurityManager sm = System.getSecurityManager();
    if (sm != null) {
        checkMemberAccess(sm, Member.DECLARED, Reflection.getCallerClass(), true);
    }
    Method method = searchMethods(privateGetDeclaredMethods(false), name, parameterTypes);
    if (method == null) {
        throw new NoSuchMethodException(methodToString(name, parameterTypes));
    }
    return getReflectionFactory().copyMethod(method);
}
```

图 10-19

```
@CallerSensitive
public Method[] getDeclaredMethods() throws SecurityException {
    SecurityManager sm = System.getSecurityManager();
    if (sm != null) {
        checkMemberAccess(sm, Member.DECLARED, Reflection.getCallerClass(), true);
    }
    return copyMethods(privateGetDeclaredMethods(false));
}
```

图 10-20

具体使用如代码 10-6 所示。

代码 10-6

```java
public class People {
    public void setName(String name) {
    }
    private int getId() {
        return 1;
    }
}


public class Student extends People{
    public void setId() {
    }
    private String getName() {
        return "张三";
    }
}


public class Test {
    public static void main(String[] args) {
        try {
            Class clazz = Class.forName("com.southwind.entity.Student");
            Method[] methods = clazz.getMethods();
```

```
            System.out.println("***************公有方法***************");
            for (Method method : methods) {
               System.out.println(method);
            }
            System.out.println("***************本类方法***************");
            Method[] declaredMethods = clazz.getDeclaredMethods();
            for (Method method : declaredMethods) {
               System.out.println(method);
            }
            System.out.println("***************公有方法 setName***************");
            Method method = clazz.getMethod("setName", String.class);
            System.out.println(method);
            System.out.println("***************本类方法 getName***************");
            Method declaredMethod = clazz.getDeclaredMethod("getName",null);
            System.out.println(declaredMethod);
         } catch (ClassNotFoundException e) {
         }
      }
   }
```

运行结果如图 10-21 所示。

```
Problems  Javadoc  Declaration  Console  Progress  Servers  Debug  Coverage
<terminated> Test (93) [Java Application] /Library/Java/JavaVirtualMachines/jdk-10.0.1.jdk/Contents/Home/bin/java (Oct 30, 2018, 4:19:36 PM)
***************公有方法***************
public void com.southwind.entity.Student.setId()
public void com.southwind.entity.People.setName(java.lang.String)
public final native void java.lang.Object.wait(long) throws java.lang.InterruptedException
public final void java.lang.Object.wait(long,int) throws java.lang.InterruptedException
public final void java.lang.Object.wait() throws java.lang.InterruptedException
public boolean java.lang.Object.equals(java.lang.Object)
public java.lang.String java.lang.Object.toString()
public native int java.lang.Object.hashCode()
public final native java.lang.Class java.lang.Object.getClass()
public final native void java.lang.Object.notify()
public final native void java.lang.Object.notifyAll()
***************本类方法***************
private java.lang.String com.southwind.entity.Student.getName()
public void com.southwind.entity.Student.setId()
***************公有方法setName***************
public void com.southwind.entity.People.setName(java.lang.String)
***************本类方法getName***************
private java.lang.String com.southwind.entity.Student.getName()
```

图 10-21

10.2.5 获取成员变量

Class 提供了 4 个方法获取类的成员变量，分别是 getFields()、getDeclaredFields()、getField(String name)和 getDeclaredField(String name)，各个方法的定义如图 10-22～图 10-25 所示。

```
@CallerSensitive
public Field[] getFields() throws SecurityException {
    SecurityManager sm = System.getSecurityManager();
    if (sm != null) {
        checkMemberAccess(sm, Member.PUBLIC, Reflection.getCallerClass(), true);
    }
    return copyFields(privateGetPublicFields());
}
```

图 10-22

```
@CallerSensitive
public Field[] getDeclaredFields() throws SecurityException {
    SecurityManager sm = System.getSecurityManager();
    if (sm != null) {
        checkMemberAccess(sm, Member.DECLARED, Reflection.getCallerClass(), true);
    }
    return copyFields(privateGetDeclaredFields(false));
}
```

图 10-23

```
@CallerSensitive
public Field getField(String name)
    throws NoSuchFieldException, SecurityException {
    Objects.requireNonNull(name);
    SecurityManager sm = System.getSecurityManager();
    if (sm != null) {
        checkMemberAccess(sm, Member.PUBLIC, Reflection.getCallerClass(), true);
    }
    Field field = getField0(name);
    if (field == null) {
        throw new NoSuchFieldException(name);
    }
    return getReflectionFactory().copyField(field);
}
```

图 10-24

```
@CallerSensitive
public Field getDeclaredField(String name)
    throws NoSuchFieldException, SecurityException {
    Objects.requireNonNull(name);
    SecurityManager sm = System.getSecurityManager();
    if (sm != null) {
        checkMemberAccess(sm, Member.DECLARED, Reflection.getCallerClass(), true);
    }
    Field field = searchFields(privateGetDeclaredFields(false), name);
    if (field == null) {
        throw new NoSuchFieldException(name);
    }
    return getReflectionFactory().copyField(field);
}
```

图 10-25

具体使用如代码 10-7 所示。

代码 10-7

```
public class People {
    public int id;
    private String name;
}


public class Student extends People{
    public int age;
    private String hobby;
}


public class Test {
    public static void main(String[] args) {
        try {
```

10.2 获取类结构

```
            Class clazz = Class.forName("com.southwind.entity.Student");
            Field[] fields = clazz.getFields();
            System.out.println("***************公有成员变量***************");
            for (Field field : fields) {
                System.out.println(field);
            }
            System.out.println("***************本类成员变量***************");
            Field[] declaredFields = clazz.getDeclaredFields();
            for (Field field : declaredFields) {
                System.out.println(field);
            }
            System.out.println("***************公有成员变量id***************");
            Field field = clazz.getField("id");
            System.out.println(field);
            System.out.println("***************本类成员变量hobby***************");
            Field declaredField = clazz.getDeclaredField("hobby");
            System.out.println(declaredField);
        } catch (ClassNotFoundException e) {
        }
    }
}
```

运行结果如图 10-26 所示。

```
***************公有成员变量***************
public int com.southwind.entity.Student.age
public int com.southwind.entity.People.id
***************本类成员变量***************
public int com.southwind.entity.Student.age
private java.lang.String com.southwind.entity.Student.hobby
***************公有成员变量id***************
public int com.southwind.entity.People.id
***************本类成员变量hobby***************
private java.lang.String com.southwind.entity.Student.hobby
```

图 10-26

10.3 反射的应用

前面的章节我们详细讲解了 Class 类的常用方法，如获取成员变量、获取方法等。那么我们获取到这些类信息到底有什么用呢？在实际开发中应该如何使用反射呢？本节我们就来学习反射的实际应用。

10.3.1 反射调用方法

常规情况下，我们需要先创建实例化对象，然后调用该对象的方法，操作的是实例化对象，如代码 10-8 所示。

代码 10-8

```java
public class Student{
    private int id;
    private String name;
    //getter、setter 方法

    public void showInfo() {
        System.out.println("学生信息");
        System.out.println("ID: "+this.id);
        System.out.println("姓名: "+this.name);
    }
    public static void main(String[] args) {
        Student student = new Student();
        student.setId(1);
        student.setName("张三");
        student.showInfo();
    }
}
```

运行结果如图 10-27 所示。

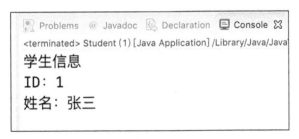

图 10-27

反射就是将常规方式进行反转，首先创建实例化对象，然后该对象的方法也抽象成对象，我们称之为方法对象。调用方法对象的 invoke 方法来实现业务需求，并将实例化对象作为参数传入 invoke 方法中。整个过程操作的是方法对象，与常规方式恰好相反，如代码 10-9 所示。

代码 10-9

```java
public class Student{
    ……
    public static void main(String[] args) {
        Student student = new Student();
        student.setId(1);
        student.setName("张三");
        Class clazz = student.getClass();
        try {
            //获取 showInfo()方法
            Method method = clazz.getDeclaredMethod("showInfo", null);
            //通过 invoke 方法完成调用
            method.invoke(student, null);
        } catch (NoSuchMethodException e) {
```

10.3 反射的应用

            }
        }
    }

运行结果如图 10-28 所示。

图 10-28

在代码 10-9 中可以看到是通过 invoke()方法来完成目标方法调用的，invoke 是 Method 类中定义的方法，该方法的作用是通过反射来执行目标方法，定义如图 10-29 所示。

```
@CallerSensitive
@ForceInline // to ensure Reflection.getCallerClass optimization
@HotSpotIntrinsicCandidate
public Object invoke(Object obj, Object... args)
    throws IllegalAccessException, IllegalArgumentException,
        InvocationTargetException
{
    if (!override) {
        Class<?> caller = Reflection.getCallerClass();
        checkAccess(caller, clazz,
                    Modifier.isStatic(modifiers) ? null : obj.getClass(),
                    modifiers);
    }
    MethodAccessor ma = methodAccessor;          // read volatile
    if (ma == null) {
        ma = acquireMethodAccessor();
    }
    return ma.invoke(obj, args);
}
```

图 10-29

该方法的参数包括两类，一类是一个 Object 类型的参数，另一类是一组 Object 类型的参数。前者表示调用该方法的对象，因为虽然是通过反射机制来执行，但是 Java 中非静态方法的调用都是由对象来完成的。后者是一组可变参数，表示调用该方法需要传入的参数。

同时该方法的返回值为 Object 类型，这里用到了多态，因为方法定义时并不知道开发者需要调用的方法返回值是什么类型，所以这里定义为 Object，可以指代任意数据类型。代码 10-8 中演示的 showInfo()方法没有返回值，接下来我们演示有返回值的 getName()方法，如代码 10-10 所示。

代码 10-10

```java
public class Student{
    ……
    public static void main(String[] args) {
        Student student = new Student();
        student.setId(1);
```

```
        student.setName("张三");
        Class clazz = student.getClass();
        try {
            //获取showInfo()方法
            Method method = clazz.getDeclaredMethod("getName", null);
            //通过invoke方法完成调用
            String name = (String) method.invoke(student, null);
            System.out.println(name);
        } catch (NoSuchMethodException e) {
        }
    }
}
```

运行结果如图 10-30 所示。

图 10-30

10.3.2 反射访问成员变量

通过反射机制，我们也可以在程序运行期间访问成员变量，并获取成员变量的相关信息，如名称、数据类型、访问权限等，具体操作如代码 10-11 所示。

代码 10-11

```
public class Student{
    private int id;
    public String name;
    //getter、setter方法
}


public class Test {
    public static void main(String[] args) {
        Class clazz = Student.class;
        Field[] fields = clazz.getDeclaredFields();
        for (Field field : fields) {
            int modifiers = field.getModifiers();
            Class typeClass = field.getType();
            String name = field.getName();
            System.out.println("成员变量"+name+"的数据类型是："+typeClass.getName()+",访问权限是："+getModifiers(modifiers));
        }
    }
    public static String getModifiers(int modifiers) {
        String result = null;
        switch (modifiers) {
        case 0:
            result = "";
```

```
            break;
        case 1:
            result = "public";
            break;
        case 2:
            result = "private";
            break;
        case 4:
            result = "protected";
            break;
        }
        return result;
    }
}
```

在代码 10-11 中，field.getModifiers()方法返回的是 int 类型的数据，不同的值对应不同的访问权限，通过 getModifiers(int modifiers)完成转换，运行结果如图 10-31 所示。

```
Problems  @ Javadoc  Declaration  Console  Progress  Servers  Debug  Coverage
<terminated> Test (95) [Java Application] /Library/Java/JavaVirtualMachines/jdk-10.0.1.jdk/Contents/Home/bin/java (Oct
成员变量id的数据类型是：int，访问权限是：private
成员变量name的数据类型是：java.lang.String，访问权限是：public
```

图 10-31

代码 10-11 完成了对成员变量的访问，获取了成员变量的信息。同时我们也可以对成员变量的值进行修改，通过调用 Field 的 set()方法即可，set()方法的定义如图 10-32 所示。

```
@CallerSensitive
@ForceInline // to ensure Reflection.getCallerClass optimization
public void set(Object obj, Object value)
    throws IllegalArgumentException, IllegalAccessException
{
    if (!override) {
        Class<?> caller = Reflection.getCallerClass();
        checkAccess(caller, obj);
    }
    getFieldAccessor(obj).set(obj, value);
}
```

图 10-32

参数列表为 Object obj 和 Object value，这两个 Object 分别表示什么呢？obj 表示被修改的对象，value 表示修改之后的值，这里的用法和上一节介绍的 Method 的 invoke()方法很相似，具体操作如代码 10-12 所示。

代码 10-12

```
public class Test {
    public static void main(String[] args) {
        Class clazz = Student.class;
        Field[] fields = clazz.getDeclaredFields();
        Student student = new Student();
        for (Field field : fields) {
            try {
```

```
                if(field.getName().equals("id")) {
                    field.set(student, 1);
                }
                if(field.getName().equals("name")) {
                    field.set(student, "张三");
                }
            } catch (IllegalArgumentException e) {
            }
        }
    }
}
```

运行程序会抛出异常，如图 10-33 所示。

```
ect.Test cannot access a member of class com.southwind.entity.Student with modifiers "private"
legalAccessException(Reflection.java:360)
eckAccess(AccessibleObject.java:589)
ield.java:1075)
a:778)
```

图 10-33

异常是因为修改了 private 修饰的成员变量，我们知道私有成员变量在外部是无法访问的，所以会抛出异常，这个问题如何解决呢？第 1 种方法可以修改成员变量的访问权限修饰符，让其变为公有的，允许外部直接访问。第 2 种方法是通过反射的"暴力修改"来解决，所谓的"暴力修改"是指修改 private 成员变量的访问权限，设置为 ture 时，表示可以修改，false 表示不能修改，会抛出异常，默认值为 false。可通过调用 setAccessible(boolean flag)方法完成权限设置，方法定义如图 10-34 所示。

```
@Override
@CallerSensitive
public void setAccessible(boolean flag) {
    AccessibleObject.checkPermission();
    if (flag) checkCanSetAccessible(Reflection.getCallerClass());
    setAccessible0(flag);
}
```

图 10-34

修改上述代码，通过"暴力修改"来设置成员变量的访问权限，如代码 10-13 所示。

代码 10-13

```
public class Test {
    public static void main(String[] args) {
        Class clazz = Student.class;
        Field[] fields = clazz.getDeclaredFields();
        Student student = new Student();
        for (Field field : fields) {
            try {
                if(field.getName().equals("id")) {
                    field.setAccessible(true);
                    field.set(student, 1);
                }
                if(field.getName().equals("name")) {
                    field.set(student, "张三");
```

```
            }
        } catch (IllegalArgumentException e) {
        }
    }
    System.out.println("学生信息");
    System.out.println("ID："+student.getId());
    System.out.println("姓名："+student.name);
    }
}
```

运行如图 10-35 所示。

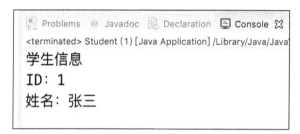

图 10-35

10.3.3　反射调用构造函数

前面的章节已经讲过，反射就是将传统的开发方式进行反转，通过对象来获取类的结构信息，并且用一个 Class 对象来描述。类的结构信息包括了成员变量、方法、构造函数等，前两节我们分别演示了反射访问成员变量以及反射调用方法，本节我们来学习通过反射机制调用构造函数创建实例化对象。

反射机制下用 Constructor 类来描述构造函数，同时一个类可以拥有多个构造函数，通过 getConstructors()方法和 getConstructor(Class<?>... parameterTypes)方法来获取相应的构造函数。前面的章节已经演示了如何获取，本节我们重点学习如何使用构造函数创建对象，通过调用 Constructor 类的 newInstance (Object ... initargs)来完成对象的创建，该方法的定义如图 10-36 所示。

```
@CallerSensitive
@ForceInline // to ensure Reflection.getCallerClass optimization
public T newInstance(Object ... initargs)
    throws InstantiationException, IllegalAccessException,
           IllegalArgumentException, InvocationTargetException
{
    if (!override) {
        Class<?> caller = Reflection.getCallerClass();
        checkAccess(caller, clazz, clazz, modifiers);
    }
    if ((clazz.getModifiers() & Modifier.ENUM) != 0)
        throw new IllegalArgumentException("Cannot reflectively create enum objects");
    ConstructorAccessor ca = constructorAccessor;   // read volatile
    if (ca == null) {
        ca = acquireConstructorAccessor();
    }
    @SuppressWarnings("unchecked")
    T inst = (T) ca.newInstance(initargs);
    return inst;
}
```

图 10-36

具体操作如代码 10-14 所示。

代码 10-14

```java
public class Student{
   private int id;
   public String name;
   //getter、setter 方法

   public Student() {
   }
   public Student(int id,String name) {
      this.id = id;
      this.name = name;
   }
   @Override
   public String toString() {
      return "Student [id=" + id + ", name=" + name + "]";
   }
}


public class Test {
   public static void main(String[] args) {
      Class clazz = Student.class;
      try {
         //获取 Student 无参构造函数
         Constructor<Student> constructor = clazz.getConstructor(null);
         Student student = constructor.newInstance(null);
         System.out.println(student);
         //获取 Student 有参构造函数
         Constructor<Student> constructor2 = clazz.getConstructor(int.class,String.class);
         Student student2 = constructor2.newInstance(1,"张三");
         System.out.println(student2);
      } catch (NoSuchMethodException e) {
      }
   }
}
```

运行如图 10-37 所示。

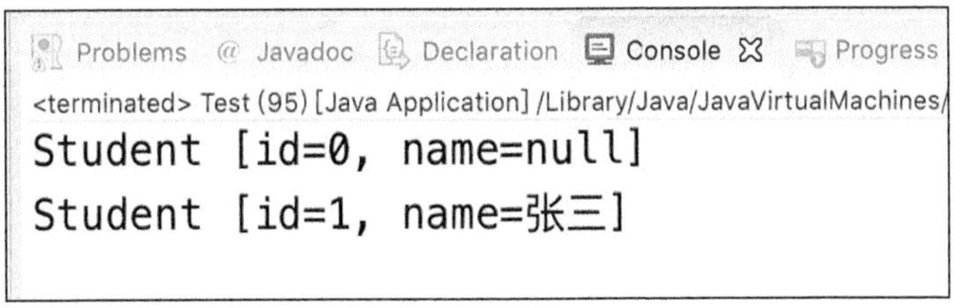

```
Student [id=0, name=null]
Student [id=1, name=张三]
```

图 10-37

10.4 动态代理

前面的章节详细讲解了反射的概念和具体操作，在实际开发中反射究竟会应用在哪些地方呢？Java 中的动态代理就是反射的一个重要应用，本节我们就来详细讲解动态代理。

代理模式是一种常用的 Java 设计模式，指的是软件设计所遵循的一套理论和准则。顾名思义，代理模式就是指在处理一个业务逻辑时，通过代理的方式完成。既然是代理的方式，就一定有被代理方或者叫委托方，和代理方，即委托方委托代理方帮他完成某些工作。代理模式的例子在日常生活中随处可见，比如你要租房子，但因为上班太忙没有大量的时间去筛选房源，这时候你就可以选择房屋中介，告诉对方你的要求，让他们帮你做筛选，然后你从中选择自己心仪的房子，完成租房需求。这就是一个代理的模式，你就是委托方，房屋中介就是代理方。在所有的代理关系中，委托方和代理方都有一个共性，即双方都具备完成需求的能力。在程序中如何表述这种关系呢？在 Java 程序中，我们把对象所具有的能力封装成接口，所以 Java 中代理模式的特点是委托类和代理类实现了同样的接口，代理类可以代替委托类完成一些核心业务以外的工作，例如消息预处理、过滤消息以及事后处理消息等。

代理类与委托类之间通过依赖注入进行关联，即在设计程序时需要将委托类定义为代理类的成员变量。代理类本身并不会去执行业务逻辑，而是通过调用委托类的方法来完成的。简单来讲，我们在访问委托对象时，是通过代理对象来间接访问的。代理模式就是通过这种间接访问的方式，为程序预留出了可处理的空间，利用此空间，在不影响核心业务的基础上可以附加其他的业务，这就是代理模式的优点。

代理模式又可以分为静态代理和动态代理，两者的区别在于静态代理需要预先写好代理类的代码，在编译期代理类的 class 文件就已经生成了。而动态代理是指在编译期并没有确定具体的代理类，在程序运行期间根据 Java 代码的指示动态地生成的方式。简单地理解静态代理是预先写好代理类，动态代理是程序运行期间动态生成代理类。很显然动态代理的方式更加灵活，可以很方便对代理类的方法进行统一的处理，而不需要逐一修改每个代理类中的方法。其中静态代理跟反射没有什么直接联系，动态代理是运用反射机制来实现的。

我们先来了解静态代理，通过一个例子来编写静态代理的代码。例如销售 iPhone 和华为两个牌子的手机，首先需要定义一个接口 Phone 表示销售手机的功能，然后定义两个实现类分别销售 iPhone 和华为手机，具体实现如代码 10-15 所示。

代码 10-15

```java
public interface Phone {
    public String salePhone();
}


public class Apple implements Phone{
    @Override
    public String salePhone() {
        // TODO Auto-generated method stub
        return "销售iPhone手机";
    }
}
```

```java
public class HuaWei implements Phone {
    @Override
    public String salePhone() {
        // TODO Auto-generated method stub
        return "销售华为手机";
    }
}
```

两个厂家销售手机的过程如代码 10-16 所示。

代码 10-16

```java
public class Test {
    public static void main(String[] args) {
        Phone phone1 = new Apple();
        System.out.println(phone1.salePhone());
        Phone phone2 = new HuaWei();
        System.out.println(phone2.salePhone());
    }
}
```

运行结果如图 10-38 所示。

图 10-38

现在有一家代理手机厂商，既可以销售 iPhone 手机也可以销售华为手机，用 Java 程序实现这一过程就需要用到静态代理模式。先创建 PhoneProxy 类，实现 Phone 接口，同时定义一个 Phone 类型的成员变量，用来接收委托对象，并代理委托对象完成 salePhone() 方法。实际是调用委托对象的 salePhone()方法，并且可以添加其他的业务逻辑，具体实现如代码 10-17 所示。

代码 10-17

```java
public class PhoneProxy implements Phone{
    private Phone phone;
    public PhoneProxy(Phone phone) {
        this.phone = phone;
    }
    @Override
    public String salePhone() {
        // TODO Auto-generated method stub
        System.out.println("代理模式");
        return this.phone.salePhone();
    }
}
```

```java
public class Test {
    public static void main(String[] args) {
        PhoneProxy phoneProxy = new PhoneProxy(new Apple());
        System.out.println(phoneProxy.salePhone());
        phoneProxy = new PhoneProxy(new HuaWei());
        System.out.println(phoneProxy.salePhone());
    }
}
```

运行结果如图 10-39 所示。

图 10-39

这样就完成了一个静态代理，其优势在于如果要完成业务扩展，不需要修改委托类 Apple 或者 HuaWei 中的 salePhone()方法，只需要改动代理类 PhoneProxy 中的 salePhone() 方法即可，这在分离不同业务的同时保证代码的整洁。同理我们来完成一个代理厂商可以同时销售宝马和奔驰两个品牌汽车的例子，具体实现如代码 10-18 所示。

代码 10-18

```java
public interface Car {
    public String saleCar();
}



public class BMW implements Car{
    @Override
    public String saleCar() {
        // TODO Auto-generated method stub
        return "销售宝马汽车";
    }
}



public class Benz implements Car{
    @Override
    public String saleCar() {
        // TODO Auto-generated method stub
        return "销售奔驰汽车";
    }
}
```

```java
public class CarProxy implements Car{
    private Car car;
    public CarProxy(Car car) {
        this.car = car;
    }
    @Override
    public String saleCar() {
        // TODO Auto-generated method stub
        System.out.println("代理模式");
        return this.car.saleCar();
    }
}

public class Test {
    public static void main(String[] args) {
        CarProxy carProxy = new CarProxy(new BMW());
        System.out.println(carProxy.saleCar());
        carProxy = new CarProxy(new Benz());
        System.out.println(carProxy.saleCar());
    }
}
```

运行结果如图 10-40 所示。

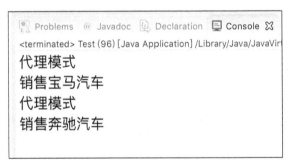

图 10-40

好了，现在的需求是创建一个厂商，既可以代理销售手机又可以代理销售汽车。那么使用静态代理肯定是不行的，因为无论是 PhoneProxy 还是 CarPhone，都只能代理一种商品的销售，要么是手机要么是汽车。然而 PhoneProxy 无法代理销售汽车，CarPhone 无法代理销售手机，因为 PhoneProxy 类和 CarPhone 类都是提前写好的，无法根据需求动态修改。

使用动态代理就可以很好地完成这一需求，前面我们提到过动态代理是指在程序运行时动态生成代理类，那么这个动态生成的功能是谁来完成的呢？java.lang.reflect 包中提供了 InvocationHandler 接口，通过该接口可以在程序运行期间动态生成代理类。首先，自定义一个类 MyInvocationHandler，实现 InvocationHandler 接口，这个类就是动态代理类的模版，如代码 10-19 所示。

代码 10-19

```java
public class MyInvocationHandler implements InvocationHandler {
    private Object obj = null;
```

```java
    public Object bind(Object obj) {
        this.obj = obj;
        return Proxy.newProxyInstance(MyInvocationHandler.class.getClassLoader(),
 obj.getClass().getInterfaces(), this);
    }
    @Override
    public Object invoke(Object proxy, Method method, Object[] args) throws Throwable {
        // TODO Auto-generated method stub
        Object result = method.invoke(obj, args);
        return result;
    }
}
```

MyInvocationHandler 类中定义的委托类成员变量的数据类型为 Object，它可以接收任意数据类型的委托类，即这个代理商什么样的产品都可以代理，这里就用到了多态的机制。MyInvocationHandler 类的 bind()方法的作用是返回一个代理对象供外部调用，这个代理对象是通过 Proxy 类的 newProxyInstance()方法来创建的，该方法定义如图 10-41 所示。

```java
@CallerSensitive
public static Object newProxyInstance(ClassLoader loader,
                                      Class<?>[] interfaces,
                                      InvocationHandler h) {
    Objects.requireNonNull(h);

    final Class<?> caller = System.getSecurityManager() == null
                                ? null
                                : Reflection.getCallerClass();

    /*
     * Look up or generate the designated proxy class and its constructor.
     */
    Constructor<?> cons = getProxyConstructor(caller, loader, interfaces);

    return newProxyInstance(caller, cons, h);
}
```

图 10-41

参数列表中的 loader 是类加载器，在程序运行期将动态生成的代理类加载到内存中。interfaces 是委托类的接口，动态代理机制需要获取到委托类的所有接口信息，以便让动态代理类也具备相同的功能，h 表示当前的 InvocationHandler 对象。在代码 10-20 中，MyInvocationHandler 类中还有一个 invoke()方法，在该方法中可以获取到委托类的方法对象 Method，然后通过反射机制来调用委托对象的业务方法，同时可以添加其他的业务逻辑代码，具体实现如代码 10-20 所示。

代码 10-20

```java
public class MyInvocationHandler implements InvocationHandler {
    ……
    @Override
    public Object invoke(Object proxy, Method method, Object[] args) throws Throwable {
        // TODO Auto-generated method stub
        System.out.println("代理模式");
        Object result = method.invoke(obj, args);
        return result;
    }
}
```

}
```

动态代理的使用如代码 10-21 所示。

代码 10-21

```
public class Test {
    public static void main(String[] args) {
        MyInvocationHandler myInvocationHandler = new MyInvocationHandler();
        Phone phone = (Phone)myInvocationHandler.bind(new Apple());
        System.out.println(phone.salePhone());
        Car car = (Car)myInvocationHandler.bind(new BMW());
        System.out.println(car.saleCar());
    }
}
```

运行结果如图 10-42 所示。

图 10-42

10.5 小结

本章为大家讲解了 Java 中的反射机制，反射的概念不好理解，较为抽象，但是它很重要，是 Java 动态语言的关键。可以说 Java 之所以很灵活，扩展性较好就是因为有反射机制，反射机制可以让程序在运行期间动态地生成某些相关模块。概念性的东西是需要通过不断地实践来验证和理解的，我们只有在实际开发中多思考，多总结才能真正体会到反射的精妙所在。本章为大家讲解了反射相关的类，以及如何通过反射机制创建实例化对象、访问成员变量和调用方法。最后讲了反射在 Java 中很重要的一个实际应用：动态代理，以此加深大家对反射的理解。
```

第 11 章 网络编程

> 如今是互联网时代，每个人的日常生活都离不开网络，Web 编程在整个 Java 开发中占据了很大一部分内容。本章我们就来学习如何用 Java 进行 Web 编程。所谓的 Web 编程就是编写程序运行在同一网络下的两个终端上，使得它们之间可以进行数据传输。在正式学习 Java Web 编程之前，我们先来了解网络的相关基础知识。

计算机网络就是通过硬件设施、传输媒介把各个不同物理地址上的计算机进行连接，形成一个资源共享和数据传输的网络系统，如图 11-1 所示。

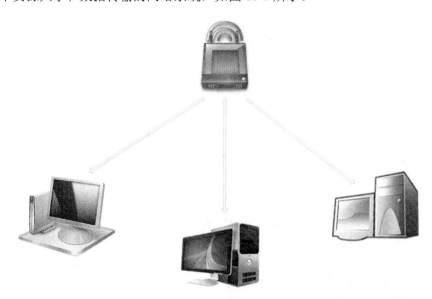

图 11-1

两台终端通过网络进行连接时，需要遵守一定的规则，这个规则就是网络协议（network rotocol），网络协议主要由 3 个特征组成。

- 语法：数据信息的结构。
- 语义：描述请求、动作和响应。
- 同步：动作的实现顺序。

网络通信协议有 TCP/IP 协议、IPX/SPX 协议、NetBEUI 协议等，我们常用的是 TCP/IP

协议，同时 TCP/IP 协议是分层的，分层的优点如下：

- 各层之间相互独立，互不干扰；
- 维护性，扩展性好；
- 有利于系统的标准化。

分层的思想在程序开发中的应用非常普遍，它的好处是每一层只关注自己的业务，无需去关注其他层的业务，只需要获取其他层传来的信息，进行处理之后再传给下一层。例如我们在编写 Java 代码的时候，不需要考虑底层的操作系统是 Windows 还是 Mac。分层思想已经帮我们屏蔽了底层机制，我们只需要在应用层写业务代码即可。TCP/IP 协议可以分为 4 层，从上到下依次为应用层、传输层、网络层和网络接口层，如图 11-2 所示。

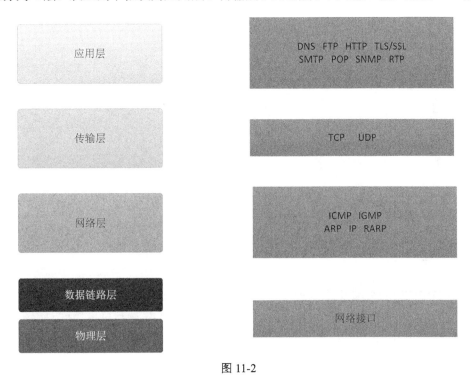

图 11-2

- 应用层（application layer）是整个体系结构中的顶层，通过应用程序之间的数据交互完成网络应用。
- 传输层（transport layer）为两台终端中应用程序之间的数据交互提供数据传输服务。
- 网络层（network layer）也叫 IP 层，负责为网络中不同的终端提供通信服务。
- 网络接口层（network interface layer）包括数据链路层（data link layer）和物理层（physical layer），数据链路层的作用是为两台终端的数据传输提供链路协议；物理层是指光纤、电缆或者电磁波等真实存在的物理媒介，这些媒介可以传送网络信号。

终端 A 正在和终端 B 通过网络进行通信，整个数据的传输流程是终端 A→应用层→传输层→网络层→数据链路层→网络层→传输层→应用层→终端 B，如图 11-3 所示。

图 11-3

接下来我们详细学习网络的具体概念。

11.1 IP 与端口

11.1.1 IP

互联网中的每台终端设备都有一个唯一标识,网络中的请求可以根据这个标识找到具体的计算机,这个唯一标识就是 IP 地址(Internet Protocol),用户可以通过操作系统的设置来查看本机的 IP 地址。IP 地址是 32 位的二进制数据,但是我们所看到的 IP 地址已经转为了十进制数据,如图 11-4 所示。

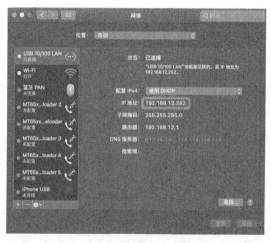

图 11-4

IP 地址 = {<网络地址>，<主机地址>}，网络地址的作用是找到主机所在的网络，主机地址的作用是找到网络中的主机。IP 地址分为 5 类，各类地址可使用的 IP 数量不同，具体范围如表 11-1 所示。

表 11-1

分类	范围
A 类	1.0.0.1 ～ 126.255.255.254
B 类	128.0.0.1 ～ 191.255.255.254
C 类	192.0.0.1 ～ 223.255.255.254
D 类	224.0.0.1 ～ 239.255.255.254
E 类	240.0.0.1 ～ 255.255.255.254

需要注意的是，我们在实际开发中并不需要记住本机的 IP 地址，可以使用 127.0.0.1 或者 localhost 来表示本机 IP 地址，Java 中有专门的类来描述 IP 地址，这个类是 java.net.InetAddress，定义如图 11-5 所示。

```
public
class InetAddress implements java.io.Serializable {
```

图 11-5

该类的常用方法如表 11-2 所示。

表 11-2

方法	描述
public static InetAddress getLocalHost() throws UnknownHostException	获取本地主机的 InetAddress 对象
public static InetAddress getByName(String host) throws UnknownHostException	通过主机名称创建 InetAddress 对象
String getHostName()	获取主机名称
public String getHostAddress()	获取主机 IP 地址
public static InetAddress getByAddress(String host, byte[] addr) throws UnknownHostException	通过主机名称和 IP 地址创建 InetAddress 对象
public static InetAddress getByAddress(byte[] addr) throws UnknownHostException	通过 IP 地址创建 InetAddress 对象

InetAddress 常用方法的使用如代码 11-1 所示。

代码 11-1

```
public class Test {
    public static void main(String[] args) {
        try {
            InetAddress inetAddress = InetAddress.getLocalHost();
            System.out.println(inetAddress.getHostName());
            System.out.println(inetAddress.getHostAddress());
            InetAddress inetAddress2 = InetAddress.getByName("127.0.0.1");
            System.out.println(inetAddress2);
            inetAddress2 = InetAddress.getByName("localhost");
```

```
            System.out.println(inetAddress2);
        } catch (UnknownHostException e) {
        }
    }
}
```

运行结果如图 11-6 所示。

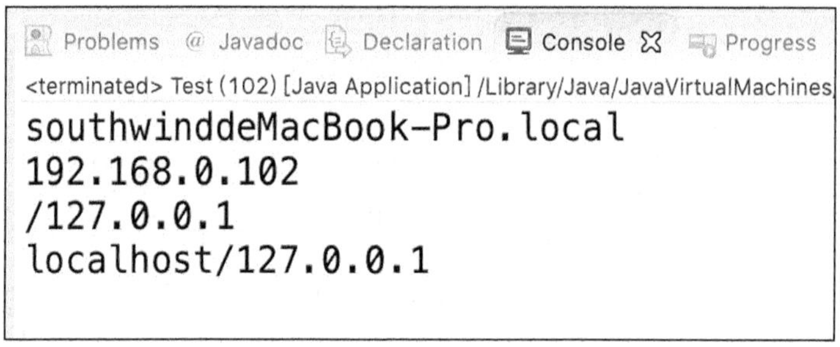

```
Problems  @ Javadoc  Declaration  Console  Progress
<terminated> Test (102) [Java Application] /Library/Java/JavaVirtualMachines
southwinddeMacBook-Pro.local
192.168.0.102
/127.0.0.1
localhost/127.0.0.1
```

图 11-6

11.1.2 端口

如果把 IP 比作一栋大厦的地址，那么端口（port）就是不同房间的门牌号。IP 地址需要和端口结合起来使用，好比快递小哥必须要通过大厦地址和房间号才能准确找到你。计算机主机就相当于大厦，网络中的请求需要通过 IP 地址来找到主机，同时一台主机上会同时运行很多个服务，如何把不同的请求分配给不同的服务就需要使用到端口了。

例如你的计算机同时打开了微信和 QQ，这是一台主机上的两个服务，朋友通过 QQ 给你发送了一条消息，那么为什么请求来到主机是被 QQ 服务所接收到而不是微信呢？就是因为不同的服务会有不同的端口，请求根据"IP 地址+端口"就可以准确地找到互联网中的接收它的服务了。比如要连接本地的 MySQL 数据库服务，MySQL 数据库服务的端口是 3306，那么完整的 URL 请求就是 localhost:3306。

11.2 URL 和 URLConnection

11.2.1 URL

通常讲的网络资源实际是指网络中真实存在的一个实体，比如文字、图片、视频、音频等，如果我们要在程序获取网络实体，应该怎么做呢？可以用 URI（Uniform Resource Identifier）统一资源定位符指向目标实体，URI 的作用是用特定的语法来标识某个网络资源。Java 中专门封装了一个类用来描述 URI，这个类就是 java.net.URI，使用 URI 的实例化对象，可以用面向对象的方式来管理网络资源，如获取主机地址、端口等，在本机（127.0.0.1）上部署服务（libmanagesys），通过 URI 读取该服务的 login.jsp 资源的具体操作如代码 11-2 所示。

代码 11-2

```
public class Test{
    public static void main(String[] args) {
        try {
            URI uri = new URI("http://localhost:8080/libmanagesys/login.jsp");
            System.out.println(uri.getHost());
            System.out.println(uri.getPort());
            System.out.println(uri.getPath());
        } catch (URISyntaxException e) {
        }
    }
}
```

运行结果如图 11-7 所示。

图 11-7

那什么是 URL 呢？URL 和 URI 有什么区别呢？URL（Uniform Resource Locator）是统一资源位置，在 URI 的基础上进行了扩充，在定位资源的同时还提供了对应的网络地址，Java 也对 URL 进行了封装，java.net.URL 类常用方法如表 11-3 所示。

表 11-3

方法	描述
public URL(String protocol, String host, int port, String file) throws MalformedURLException	根据协议、IP 地址、端口号、资源名称获取 URL 对象
public final InputStream openStream() throws java.io.IOException	获取输入流对象

在本机（127.0.0.1）上部署服务（libmanagesys），使用 URL 读取该服务的 login.jsp 资源如代码 11-3 所示。

代码 11-3

```
public class Test {
    public static void main(String[] args) {
        InputStream inputStream = null;
        Reader reader = null;
        BufferedReader bufferedReader = null;
        try {
            URL url = new URL("http","127.0.0.1",8080,"/libmanagesys/login.jsp");
            inputStream = url.openStream();
            reader = new InputStreamReader(inputStream);
            bufferedReader = new BufferedReader(reader);
```

```
                String str = null;
                while ((str = bufferedReader.readLine())!=null) {
                    System.out.println(str);
                }
        } catch (MalformedURLException e) {
        } finally {
            try {
                bufferedReader.close();
                reader.close();
                inputStream.close();
            } catch (IOException e) {
            }
        }
    }
}
```

运行结果如图 11-8 所示。

```
<html>
<head>
    <title></title>
    <link rel="stylesheet" href="layui/css/layui.css"  media="all">
</head>
<body>
<div class="layui-container" style="width: 500px;height: 330px;margin-top
    <form class="layui-form" action="login.do" method="post">
        <div class="layui-form-item">
            <label class="layui-form-label">用户名：</label>
            <div class="layui-inline">
                <input type="text" name="username" lay-verify="username"
            </div>
        </div>
```

图 11-8

同时也可以通过 URL 获取资源的其他信息，如主机地址、端口等，具体操作如代码 11-4 所示。

代码 11-4

```
public class Test {
    public static void main(String[] args) {
        InputStream inputStream = null;
        Reader reader = null;
        BufferedReader bufferedReader = null;
        try {
            URL url = new URL("http","127.0.0.1",8080,"/libmanagesys/login.jsp");
            System.out.println(url.getHost());
            System.out.println(url.getPort());
            System.out.println(url.getPath());
        } catch (MalformedURLException e) {
        }
    }
}
```

运行结果如图 11-9 所示。

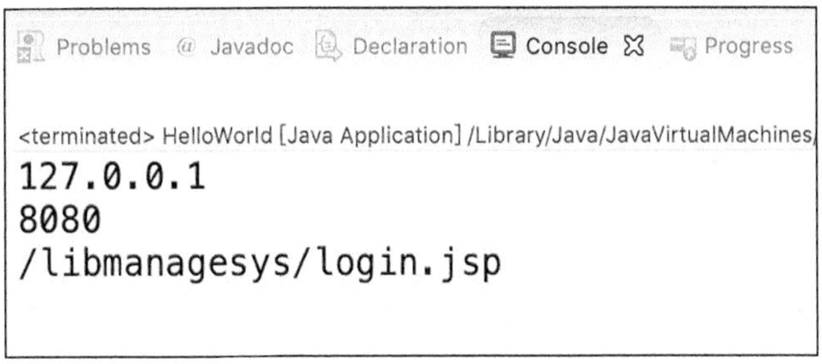

Problems @ Javadoc Declaration Console Progress
<terminated> HelloWorld [Java Application] /Library/Java/JavaVirtualMachines/
127.0.0.1
8080
/libmanagesys/login.jsp

图 11-9

11.2.2 URLConnection

URLConnection 用来描述 URL 指定资源的连接，是一个抽象类，常用的子类有 HttpURLConnection，URLConnection 底层是通过 HTTP 协议来处理的，它定义了访问远程网络资源的方法。通过 URLConnection 可以获取到 URL 资源的相关信息，该类常用的方法如表 11-4 所示。

表 11-4

方法	描述
public int getContentLength()	返回资源的长度，返回值为 int 类型
public long getContentLengthLong()	返回资源的长度，返回值为 long 类型
public String getContentType()	返回资源的类型
public abstract void connect() throws IOException	判断连接的打开或关闭状态
public Inputstream getInputStream()throws IOException	获取输入流对象

接下来我们学习 URLConnection 如何使用，例如要获取代码 11-3 示例中的 URL 资源的相关信息和具体内容，就可以通过 URLConnection 来完成，具体操作如代码 11-5 所示。

代码 11-5

```
public class Test {
   public static void main(String[] args) {
      try {
         URL url = new URL("http://localhost:8080/libmanagesys/login.jsp");
         URLConnection urlConnection = url.openConnection();
         System.out.println(urlConnection.getURL());
         System.out.println(urlConnection.getContentLengthLong());
         System.out.println(urlConnection.getContentLength());
         System.out.println(urlConnection.getContentType());
         InputStream inputStream = urlConnection.getInputStream();
         Reader reader = new InputStreamReader(inputStream);
         BufferedReader bufferedReader = new BufferedReader(reader);
         String str = null;
         while ((str = bufferedReader.readLine())!=null) {
         System.out.println(str);
         }
      } catch (MalformedURLException e) {
```

```
            }
        }
    }
```

运行结果如图 11-10 所示。

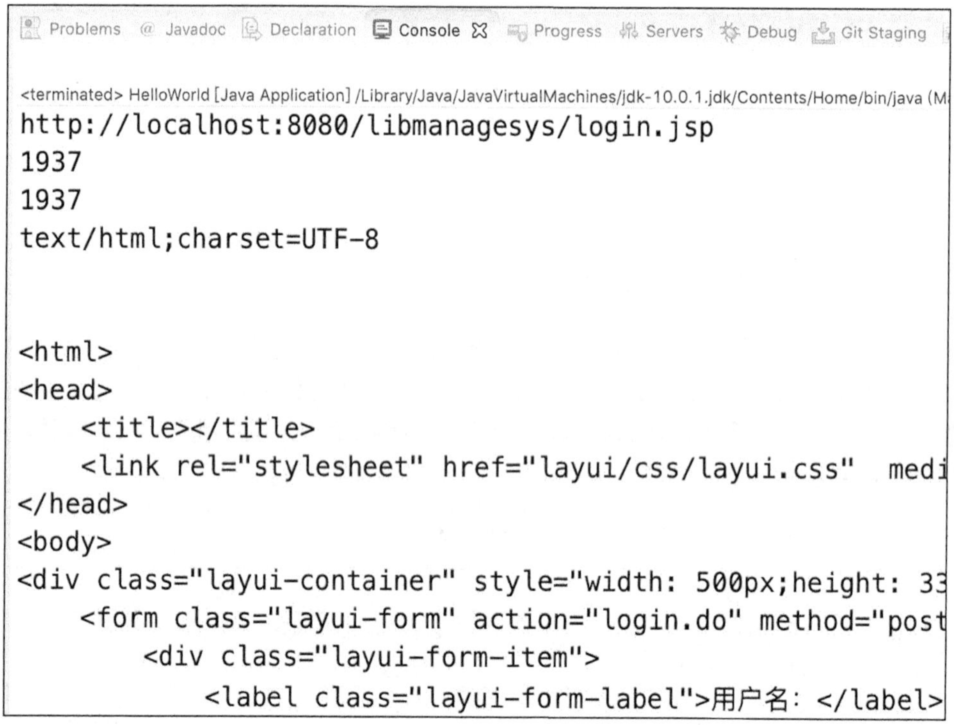
Problems
@ Javadoc
Declaration
Console
Progress
Servers
Debug
Git Staging
<terminated> HelloWorld [Java Application] /Library/Java/JavaVirtualMachines/jdk-10.0.1.jdk/Contents/Home/bin/java (M
http://localhost:8080/libmanagesys/login.jsp
1937
1937
text/html;charset=UTF-8
<html>
<head>
<title></title>
<link rel="stylesheet" href="layui/css/layui.css" medi
</head>
<body>
<div class="layui-container" style="width: 500px;height: 33
<form class="layui-form" action="login.do" method="post
<div class="layui-form-item">
<label class="layui-form-label">用户名: </label>

图 11-10

11.3 TCP 协议

TCP 是面向连接的运输层协议，比较复杂，应用程序在使用 TCP 协议前必须先建立连接，才能传输数据，数据传输完毕之后需要释放已经建立的连接。这相当于我们日常生活中的打电话，首先确保电话已经打通，才能进行对话，如果电话没有打通，就一直处于阻塞状态，需要持续等待。TCP 的优点是非常可靠，通过 TCP 传输的数据，不会出现丢失的情况，并且数据是按照先后顺序依次到达的。缺点是速度慢、效率低，实际开发中需要根据具体的业务需求来选择，对安全性要求较高的系统需要使用 TCP 协议（例如金融系统），必须先确保用户成功登录，才能进行后续的操作。

Java 通过 Socket 来完成 TCP 程序的开发，Socket 是一个类，使用该类可以在服务端与客户端之间建立可靠的连接。在实际开发中，Socket 表示客户端，服务端使用 ServerSocket 来表示，ServerSocket 也是一个类，ServerSocket 和 Socket 都存放在 java.net 包中。具体的开发思路是在服务端创建 ServerSocket 对象，然后通过该对象的 accept()方法可以接收到若干个表示客户端的 Sokcet 对象，如图 11-11 所示。

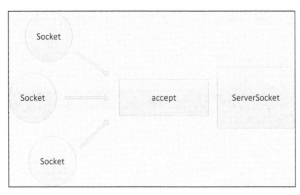

图 11-11

ServerSocket 类的常用方法如表 11-5 所示。

表 11-5

方法	描述
public ServerSocket(int port) throws IOException	根据端口创建 ServerSocke 实例对象
public ServerSocket(int port, int backlog) throws IOException	根据端口和 backlog 创建 ServerSocket 实例对象
public ServerSocket(int port, int backlog, InetAddress address) throws IOException	根据端口、backlog 和 IP 地址创建 ServerSocket 实例对象
public ServerSocket() throws IOException	创建没有绑定服务器的 ServerSocket 实例对象
public synchronized int getSoTimeout() throws IOException	获取 Sotimeout 的设置
public InetAddress getInetAddress()	获取服务器的 IP 地址
public Socket accept() throws IOException	等待客户端请求,并返回 Socket 对象
public void close() throws IOException	关闭 ServerSocket
public boolean isClosed()	返回 ServerSocket 的关闭状态
public void bind(SocketAddress endpoint) throws IOException	将 ServerSocket 实例对象绑定到指定地址
public int getLocalPort()	返回 ServerSocket 的端口

Socket 类的常用方法如表 11-6 所示。

表 11-6

方法	描述
public Socket(String host, int port) throws UnknownHostException, IOException	根据主机、端口创建要连接的 Socket 对象
public Socket(InetAddress host, int port) throws IOException	根据 IP 地址、端口创建要连接的 Socket 对象
public Socket(String host, int port, InetAddress localAddress, int localPort) throws IOException	根据主机、端口创建要连接的 Socket 对象并将其连接到指定远程主机上的指定端口
public Socket(InetAddress host, int port, InetAddress localAddress, int localPort) throws IOException	根据主机、端口创建要连接的 Socket 对象并将其连接到指定远程地址上的指定端口
public Socket()	创建没有连接的 Socket 对象
public InputStream getInputStream() throws IOException	返回 Socket 的输入流
public synchronized void close() throws IOException	关闭 Socket
public boolean isClosed()	返回 Socket 的关闭状态

下面来看 ServerSocket 和 Socket 的实际应用，首先启动 ServerSocket，等待接收客户端请求。当接收到客户端请求后，打印"接收到了客户端请求"信息，同时向客户端返回"Hello World"，服务端代码如代码 11-6 所示。

代码 11-6

```java
public class Server {
    public static void main(String[] args) {
        ServerSocket serverSocket = null;
        Socket socket = null;
        OutputStream outputStream = null;
        InputStream inputStream = null;
        DataInputStream dataInputStream = null;
        DataOutputStream dataOutputStream = null;
        try {
            serverSocket = new ServerSocket(8080);
            System.out.println("------服务端------");
            System.out.println("已启动，等待接收客户端请求...");
            while(true) {
                socket = serverSocket.accept();
                inputStream = socket.getInputStream();
                dataInputStream = new DataInputStream(inputStream);
                String request = dataInputStream.readUTF();
                System.out.println("接收到了客户端请求："+request);
                outputStream = socket.getOutputStream();
                dataOutputStream = new DataOutputStream(outputStream);
                String response = "Hello World";
                dataOutputStream.writeUTF(response);
                System.out.println("给客户端作出响应："+response);
            }
        } catch (IOException e) {
        }finally {
            try {
                dataOutputStream.close();
                outputStream.close();
                dataInputStream.close();
                inputStream.close();
                socket.close();
                serverSocket.close();
            } catch (IOException e) {
            }
        }
    }
}
```

运行结果如图 11-12 所示。

图 11-12

此时程序并没有结束,在等待客户端的连接请求。我们写好客户端代码并运行,如代码 11-7 所示。

代码 11-7

```java
public class Client {
    public static void main(String[] args) {
        Socket socket = null;
        OutputStream outputStream = null;
        DataOutputStream dataOutputStream = null;
        InputStream inputStream = null;
        DataInputStream dataInputStream = null;
        try {
            socket = new Socket("127.0.0.1", 8080);
            System.out.println("------客户端------");
            String request = "你好! ";
            System.out.println("客户端说:"+request);
            outputStream = socket.getOutputStream();
            dataOutputStream = new DataOutputStream(outputStream);
            dataOutputStream.writeUTF(request);
            inputStream = socket.getInputStream();
            dataInputStream = new DataInputStream(inputStream);
            String response = dataInputStream.readUTF();
            System.out.println("服务端响应为:"+response);
        } catch (IOException e) {
        }finally {
            try {
                dataInputStream.close();
                inputStream.close();
                dataOutputStream.close();
                outputStream.close();
                socket.close();
            } catch (IOException e) {
            }
        }
    }
}
```

运行程序后,服务端代码的运行结果如图 11-13 所示。

图 11-13

客户端的运行结果如图 11-14 所示。

```
Problems   @ Javadoc   Declaration   Console ⊠   Prog
<terminated> Client (2) [Java Application] /Library/Java/JavaVirtualMac
------客户端------
客户端说：你好！
服务端响应为：Hello World
```

图 11-14

11.4　UDP 协议

　　TCP 协议可以建立稳定可靠的连接，保证数据的完整。但是 TCP 协议的缺点也很明显，先建立可靠连接，再进行操作的方式必然会造成系统运行效率低下。在实际开发中，某些业务场景对系统的运行效率要求较高，使用 TCP 协议很显然就不合适了，这时候就需要使用另外一种传输协议——UDP。

　　UDP 所有的连接都是不可靠的，即不需要建立连接，直接发送数据即可。例如通过微信聊天，发送方只需要把信息发出去，接收方可能因为网络差或者其他原因没有收到信息，但是发送方并不会因为接收方无法接收到信息而等待，它可以连续发送数据。所以 UDP 的速度更快，但是可能会造成数据丢失，安全性不高，追求速度的应用可以选择 UDP。例如语音聊天或者视频聊天，对于这类应用流畅性更重要，偶尔丢失几个数据包并不会有太大影响。Java 提供了 DatagramSocket 类和 DatagramPacket 类，来帮助开发者编写基于 UDP 协议的程序，DatagramSocket 类的常用方法如表 11-7 所示。

表 11-7

方法	描述
public DatagramSocket(int port) throws SocketException	根据端口创建 DatagramSocket 实例对象
public void send(DatagramPacket p) throws IOException	发送数据报
public synchronized void receive(DatagramPacket p) throws IOException	接收数据报
public InetAddress getInetAddress()	获取 DatagramSocket 对应的 InetAddress 对象
public boolean isConnected()	判断是否连接到服务

　　DatagramPacket 类的常用方法如表 11-8 所示。

表 11-8

方法	描述
public DatagramPacket(byte buf[], int length, InetAddress address, int port)	根据发送的数据、数据长度、IP 地址、端口，创建 DatagramPacket 实例对象
public synchronized byte[] getData()	获取接收的数据
public synchronized int getLength()	获取数据长度
public synchronized int getPort()	获取发送数据的 Socket 端口
public synchronized SocketAddress getSocketAddress()	获取发送数据的 Socket 信息

两个终端通过 UDP 协议进行通信的具体实现如代码 11-8 所示。

代码 11-8

```java
public class TerminalA {
    public static void main(String[] args) throws Exception {
        //接收数据
        byte[] buff = new byte[1024];
        DatagramPacket datagramPacket = new DatagramPacket(buff,buff.length);
        DatagramSocket datagramSocket = new DatagramSocket(8181);
        datagramSocket.receive(datagramPacket);
        String mess = new String(datagramPacket.getData(),0,datagramPacket.getLength());
        System.out.println("我是 TerminalA,接收到了"+datagramPacket.getPort()+"传来的数据:"+mess);
        //发送数据
        String reply = "我是 TerminalA,已接收到你发来的数据。";
        SocketAddress socketAddress = datagramPacket.getSocketAddress();
        DatagramPacket datagramPacket2 = new DatagramPacket(reply.getBytes(),reply.getBytes().length,socketAddress);
        datagramSocket.send(datagramPacket2);
    }
}


public class TerminalB {
    public static void main(String[] args) throws Exception {
        String mess = "我是 TerminalB,你好! ";
        //发送数据
        InetAddress inetAddress = InetAddress.getByName("localhost");
        DatagramPacket datagramPacket = new DatagramPacket(mess.getBytes(),mess.getBytes().length,inetAddress,8181);
        DatagramSocket datagramSocket = new DatagramSocket(8080);
        datagramSocket.send(datagramPacket);
        //接收数据
        byte[] buff = new byte[1024];
        DatagramPacket datagramPacket2 = new DatagramPacket(buff,buff.length);
        datagramSocket.receive(datagramPacket2);
        String reply = new String(datagramPacket2.getData(),0,datagramPacket2.getLength());
        System.out.println("我是 TerminalB,接收到了"+datagramPacket2.getPort()+"返回的数据:"+reply);
    }
}
```

运行程序后,TerminalA 的结果如图 11-15 所示。

```
我是TerminalA, 接收到了8080传来的数据: 我是TerminalB, 你好!
```

图 11-15

运行程序后,TerminalB 的结果如图 11-16 所示。

```
Problems  Javadoc  Declaration  Console ⊠  Progress  Servers  Debug  Git Staging  Coverage
<terminated> TerminalB [Java Application] /Library/Java/JavaVirtualMachines/jdk-10.0.1.jdk/Contents/Home/bin/java (Dec 28, 2018, 11:12
我是TerminalB，接收到了8181返回的数据：我是TerminalA，已接收到你发来的数据。
```

图 11-16

11.5 多线程下的网络编程

前面章节的代码都是基于单点连接的方式，即一个服务端对应一个客户端。实际运行环境中是一个服务端需要对应多个客户端的，这种情况我们可以使用多线程来模拟，具体实现如代码 11-9 所示。

代码 11-9

```java
public class ServerThread {
    public static void main(String[] args) {
        ServerSocket serverSocket = null;
        try {
            serverSocket = new ServerSocket(8080);
            System.out.println("服务器已启动...");
            while(true) {
                Socket socket = serverSocket.accept();
                new Thread(new ServerRunnable(socket)).start();
            }
        } catch (IOException e) {
        }
    }
}

class ServerRunnable implements Runnable{
    private Socket socket;
    public ServerRunnable(Socket socket) {
        this.socket = socket;
    }
    @Override
    public void run() {
        InputStream inputStream = null;
        DataInputStream dataInputStream = null;
        try {
            inputStream = this.socket.getInputStream();
            dataInputStream = new DataInputStream(inputStream);
            String message = dataInputStream.readUTF();
            System.out.println(message);
        } catch (IOException e) {
        }finally {
            try {
                dataInputStream.close();
                inputStream.close();
            } catch (IOException e) {
            }
```

```java
        }
    }
}

public class ClientThread {
    public static void main(String[] args) {
        for(int i = 0; i < 100; i++) {
            new Thread(new ClientRunnable(i)).start();
        }
    }
}

class ClientRunnable implements Runnable{
    private int num;
    public ClientRunnable(int num) {
        this.num = num;
    }
    @Override
    public void run() {
        Socket socket = null;
        OutputStream outputStream = null;
        DataOutputStream dataOutputStream = null;
        try {
            socket = new Socket("localhost", 8080);
            String mess = "我是客户端"+this.num;
            outputStream = socket.getOutputStream();
            dataOutputStream = new DataOutputStream(outputStream);
            dataOutputStream.writeUTF(mess);
        } catch (IOException e) {
            // TODO Auto-generated catch block
            e.printStackTrace();
        }finally {
            try {
                dataOutputStream.close();
                outputStream.close();
                socket.close();
            } catch (IOException e) {
            }
        }
    }
}
```

运行 ServerThread，结果如图 11-17 所示。

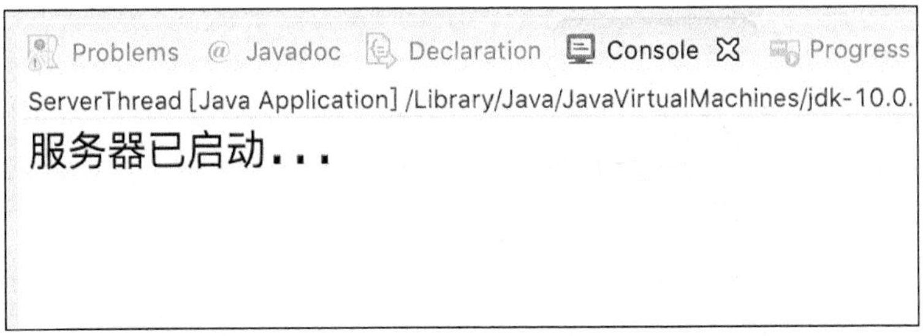

图 11-17

运行 ClientThread，结果如图 11-18 所示。

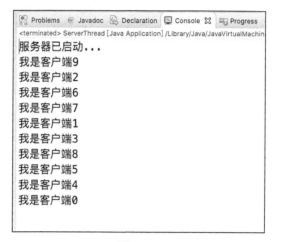

图 11-18

11.6 综合练习

使用 Socket 和多线程编写一个简单的聊天小程序，要求客户端和服务端交替发送消息，在客户端和服务端都能看到彼此的聊天记录，如图 11-19 和图 11-20 所示。

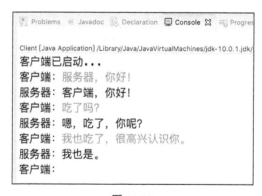

图 11-19

图 11-20

具体实现如代码 11-10 所示。

代码 11-10

```java
public class Server {
    public static void main(String[] args) {
        ServerSocket serverSocket = null;
        Socket socket = null;
        try {
            serverSocket = new ServerSocket(8080);
            System.out.println("服务器已启动...");
            while(true) {
                socket = serverSocket.accept();
                new Thread(new SocketThread(socket)).start();
            }
        } catch (IOException e) {
            // TODO Auto-generated catch block
            e.printStackTrace();
        }
    }
}


public class SocketThread implements Runnable{
    private Socket socket;
    public SocketThread(Socket socket) {
        this.socket = socket;
    }
@Override
public void run() {
    // TODO Auto-generated method stub
    InputStream inputStream = null;
    DataInputStream dataInputStream = null;
    Scanner scanner = new Scanner(System.in);
    String message = null;
    try {
        while(true) {
            //读
            inputStream = socket.getInputStream();
            dataInputStream = new DataInputStream(inputStream);
            message = dataInputStream.readUTF();
            System.out.println("客户端: "+message);

            //写
            System.out.print("服务器: ");
            message = scanner.next();
            OutputStream outputStream = null;
            DataOutputStream dataOutputStream = null;
            try {
                outputStream = socket.getOutputStream();
                dataOutputStream = new DataOutputStream(outputStream);
                dataOutputStream.writeUTF(message);
            } catch (IOException e) {
            }
        }
    } catch (IOException e) {
```

```java
            }
        }
    }

public class Client {
    public static void main(String[] args) {
        Socket socket = null;
        InputStream inputStream = null;
        DataInputStream dataInputStream = null;
        try {
            System.out.println("客户端已启动...");
            socket = new Socket("127.0.0.1", 8080);
            Scanner scanner = new Scanner(System.in);
            String message = null;
            while(true) {
                //写
                System.out.print("客户端：");
                message = scanner.next();
                OutputStream outputStream = null;
                DataOutputStream dataOutputStream = null;
                try {
                    outputStream = socket.getOutputStream();
                    dataOutputStream = new DataOutputStream(outputStream);
                    dataOutputStream.writeUTF(message);
                } catch (IOException e) {
                    // TODO Auto-generated catch block
                    e.printStackTrace();
                }

                //读
                inputStream = socket.getInputStream();
                dataInputStream = new DataInputStream(inputStream);
                message = dataInputStream.readUTF();
                System.out.println("服务器："+message);
            }
        } catch (IOException e) {
        }
    }
}
```

11.7 小结

  本章为大家讲解了 Java 网络编程的相关概念，网络编程在整个 Java 开发体系中有承上启下的作用。之前我们所讲的知识都是基于单机版的程序，也就是说写出来的代码只能在本地运行。学完本章知识，我们就从单机时代进入到了网络时代，通过网络编程可以让两台终端通过程序进行数据交互。在真正的 Java Web 开发中并不会直接使用 Socket 进行编程，JavaEE 提供了相应的 Web 组件对 Socket 底层代码进行了封装，开发者只需要在应用层通过 Web 组件进行开发即可，不需要关注底层，只需要关注业务逻辑，从而提高开发效率。本章内容让大家对网络编程的底层原理有一个简单的理解，为 Java Web 开发打好基础。